04/03/15

THE

THREE-MINUTE

THE THREE-MINUTE OUTDOORSMAN

WILD SCIENCE FROM MAGNETIC DEER TO MUMBLING CARP

Robert M. Zink

University of Minnesota Press
Minneapolis • London

Some essays in this book were previously published in *Outdoor News* and *American Waterfowler*.

Photograph on page 205 by Gunnar Ries.

Photographs on pages 1, 67, 113, and 157 by U.S. Fish and Wildlife Service.

Photograph and map on pages 162–63 from Storrs L. Olson, Horace Loftin, and Steve Goodwin, "Biological, Geographical, and Cultural Origins of the Loon Hunting Tradition in Carteret County, North Carolina," *The Wilson Journal of Ornithology* 122, no. 4 (2010): 716–24; reprinted courtesy of *The Wilson Journal of Ornithology*.

Published by the University of Minnesota Press
111 Third Avenue South, Suite 290
Minneapolis, MN 55401–2520
http://www.upress.umn.edu

Design and production by Mighty Media, Inc.
Interior and text design by Chris Long

A Cataloging-in-Publication record for this book is available from the Library of Congress.

Printed in the United States of America on acid-free paper

The University of Minnesota is an equal-opportunity educator and employer.

20 19 18 17 16 15 14   10 9 8 7 6 5 4 3 2 1

CONTENTS

Preface   ix

ALL THINGS DEER

1. A Short History of Deer in North America   2
2. A Message from Our Native Birds:
   Deer Hunters Needed   8
3. The Science of Chronic Wasting Disease and Its
   Relevance for Management of White-Tailed Deer   11
4. Urban Deer: Hunting versus Birth Control   23
5. It's Taken Centuries, but We Now Know Why
   Deer Don't Ask to Use Your Compass   28
6. Why Are Medical Researchers Interested in Antlers?   31
7. Isn't It Obvious Why Deer Have Antlers?   35
8. A New Kind of (Un)Natural Selection on
   Deer Antlers: Hunting   39
9. My Deer Doctor: Take Two Acorns and
   Call Me in the Morning   42
10. Trying to Outfox Deer Ticks and Lyme Disease   47
11. Deer and Their Subspecies: Fact or Fiction?   52
12. Can Game Managers Control the Number of Deer?   56
13. Mountain Lions, Prions, and Sick Deer   60
14. The Rut (Maybe More Than You Wanted to Know)   63

IN THE WOODS

15. Hunting Spots for Wild Turkeys at the Last
    Glacial Maximum   68
16. Wolves, Coyotes, and Deer   72
17. Lead, Lead, Everywhere?   75

18. Politics and the Lead Ammo Debate   78

19. Getting the Lead Out (of Chukars)   82

20. Sounding the Alarm, Mourning Dove Style   86

21. Moaning Moose and Topi Lies   89

22. Turkeys and Love: What's Actually Happening Out There in Spring?   92

23. Looking Back at Turkey Season: What You Might Not Have Seen   95

24. When Black Bears Attack!   97

25. I Wouldn't Have Seen It If I Hadn't Believed It: A Look at the Ivory-billed Woodpecker Controversy   101

26. Recent Developments in the Climate Change News   104

27. Night of the Dead Birds, or Too Much Hitchcock?   108

28. Eagle Attacks Toddler! Then Again, Maybe Not   111

IN THE WATER

29. Recreational Fishing Alters Fish Evolution   114

30. Duck Hunting in the Low Country; or, How's Your Kooikerhondje?   117

31. Predators and Ducklings in the North Dakota Prairies   120

32. Long-Term Sexual Tensions between Male and Female Ducks   124

33. Vigilance in Ducks: More Than Meets the Eye(lid)   128

34. What Little We Knew about the Labrador Duck Just Got Littler   132

35. Mumbling Along: Lessons from the Past about Stopping the Spread of Exotic Species   134

36. What You Don't See under Your Boat   139

37. Never Be a Baby Bird   141

38. Oh, No! Duck Hunting Videos Might Not Be Realistic!   145

39. Snow Geese and Polar Bears: Collision Course?   148

40. Species Conservation at the State Level: A Fish-Eye View   152

ANIMALS AND US

41. Reconsider Your Walk with Fido?   158
42. Loon Hunting: A Bygone Tradition   161
43. Market Hunting and the Demise of the Eskimo Curlew   165
44. The Ethics of Baiting and High-Fence Ranch Hunting:
    A Perennial Debate   169
45. Hunters and Conservationists at Odds over
    Shooting Shorebirds   177
46. A Conversation about Hunting in the Netherlands   180
47. Back from the Dead: Mother Goose Goes to the Poor
    House, Cooked   183
48. Cats Outdoors and Native Birds: An Unnatural Mix   188
49. Five Million U.S. Residents Don't See the Problem
    with Their Cat Killing Just One Bird a Day   191
50. Cats on Birds: A More Insidious Side   195
51. Some We Love, Others Not So Much   197
52. RICO, the Circus, and Conflicts between
    Hunters and Nonhunters   200

ANIMAL INTELLIGENCE

53. A New Respect for Porcupine Quills   206
54. Outfoxed Again: Foxes Use Built-In Range Finders!   209
55. How Do Ground-Nesting Grouse Ever
    Breed Successfully? An Oily Subject   212
56. Our Chickadees Are Smarter Than Theirs   216
57. Neck-Deep in Guano: A Recent History of
    Chimney Swifts   218
58. Shake, Rattle, and Spray, Doggie Style   222
59. Drahthaar Follies   225
60. "Trash Birds," the Law, and Amazing Biology   227
61. The Dating Game, Antelope Style   231
62. Camouflage: One of Life's Universals   234
63. One More Cup of Coffee   237

Postscript: Confessions of a Three-Minute Outdoorsman   241

PREFACE

"Your new hunting vest looks, umm, great," my friend remarked as we stood next to his truck in the predawn light outside my house. After a long absence from Minnesota in pursuit of becoming a university professor who studies birds, I was about to embark on my first real grouse hunt. "However, you can't wear it," he said. "Your vest is camo, and we have to wear blaze orange in the field." Super. Feeling stupid is always a great way to begin a trip. Based on my knowledge of birds, I had thought that trying to sneak up on a grouse—which has color vision, even likely in the UV range—would demand an environment-matching vest. I was wrong. I recognized that I might have a lot of "book knowledge" gained from reading scientific papers, but some important points were obviously lost in translation. There were very smart people who knew just as much as I did, but they knew different things. Later I would learn that my vest worked well for duck hunting so the purchase wasn't wasted. But for the time being, a borrowed blaze-orange vest gave me access into the unbelievably fascinating world of grouse hunting. An ongoing journey in the parallel worlds of academia and the outdoors began.

It wasn't just grouse hunting. I became immersed in hunting turkeys, ducks, geese, deer, fishing, and so on. I took hunting trips to Texas and Africa. As I continued my "day job" as a scientist studying birds, I appreciated more and more the gulf between the book learners and those whose knowledge comes directly from the outdoors. True, scientists often venture afield, even conducting experiments. But sometimes they find it challenging to connect with people who have learned a lot just being outside. Or they don't know what nonscientists would find interesting, and lacking their backgrounds, it doesn't occur to them that a particular finding would be an eye-opener. I found myself in a position of having a lot of experience outdoors, and I saw the scientific articles I read in a new, personal light. I started to write about science.

My first article was almost tongue-in-cheek. A paper was published that described how some colleagues had isolated DNA from the inside of preserved eggshells of the extinct Labrador duck. Their reason was not simply because they could do it, but they wanted to confirm that the eggs were correctly identified, as no one knew for sure what the eggs of this species of duck really looked like. By comparing the DNA from the eggshells with DNA isolated from old skins of the Labrador duck, they found that none of the eggs were from Labrador ducks. I thought this was hilarious, as one museum had spent a substantial amount of money buying the only known clutch of Labrador duck eggs, and now, a hundred years later, "forensic science" revealed the ruse.

I almost chuckled as I wrote the piece and sent it to the editor of *Outdoor News* (ON). I wrote in my e-mail to the editor, "Here is a piece that, frankly, will probably not be of interest for ON, but I thought I'd send it anyway." I was not ready for his response: "Not of interest to ON? What are you talking about?! I'd love to print this." In the early days, I had not yet appreciated that I might be a conduit from the scientific side to the lay public. I now have accumulated many life experiences that not all scientists have (though many do, for sure).

As a professional ornithologist, I have published many scientific articles. These articles are not always greeted with enthusiasm by my colleagues—we often disagree with each other. I was unprepared for the generally positive response from my popular articles. In fact, I wish I got as many nice letters about my scientific articles as I have about my popular ones. I found that I really enjoy taking a morning to read an article that I think will be of general interest, do a little background research, then write about it. There are few essays in this book that did not teach me something as I worked on them—things that just make me pause, shake my head, and say to myself, wow, how cool nature can be.

My main goal became writing fun-to-read essays for people who hunt, fish, or are interested in biology and the outdoors. I hope that nonhunters will find the essays interesting. Many are summaries of papers from scientific journals, written by scientists for scientists; I think that nonscientists ought to be able to learn about the findings as well. I tried to translate these research papers into readable units, requiring just a few minutes each, from the perspective of someone like me, who hunts, fishes, and enjoys being outside but is a research scientist in his day job. I recommend that your copy of this book remain in places where you have a few minutes of time for reading, but propriety prevents me from suggesting exact locations.

I am indebted to Rob Drieslein, editor of *Outdoor News* (Minnesota) for encouraging me to write and providing many constructive ideas. Several of these articles first appeared in the pages of *Outdoor News*, and I appreciate the permission to reprint them here.

I thank Craig Johnston for setting me straight about hunting vests and inviting me to hunt ducks and geese; David Andersen for an introduction to Canada, woodcock, and Drahthaars; Al and Gerry Johnson for their hospitality in Manitoba; Don Alstad for getting me addicted to archery; the guys at Wild Wings of Oneka for a never-ending good time; the late Bud Tordoff for introducing me to English setters; and Jake Beard for early experiences. Brian Fowler fueled my addiction to sporting clays, bird hunting, and fishing the St. Croix River. Glenn and Brad have accommodated my passions by providing access to their properties. My friend Gene Wegleitner, whom I might talk to more than to my wife during the bow season, taught me some of the most important things I have learned. My sons have been more than worthy companions in all of my outdoors addictions, and my wife, Susan Weller, has gone above and beyond in her support.

To all of you who look forward to the sun rising over your duck marsh, hiking in the north woods, the last few minutes of afternoon light while in your deer stand, canoeing in the Boundary Waters, the tug on the end of your line, the sounds of a spring gobbler, and a love for learning cool things about our natural world and the animals that share it with us, I hope you find these essays interesting and fun to read.

ALL THINGS DEER

1

A SHORT HISTORY OF DEER
IN NORTH AMERICA

Today deer are just about everywhere—whether glimpsed in the woods, eating garden plants, or wrapped around your bumper—but it isn't clear they were always so populous, as historical records about deer numbers are not very reliable.

Historical records are not accurate, because they are anecdotal accounts, not recorded observations of numbers of deer. The science of wildlife management didn't even exist until the early twentieth century. A book chapter titled "Management History" by Kip Adams and Joseph Hamilton (in *Biology and Management of White-Tailed Deer*, ed. D. G. Hewitt, CRC Press, 2011) provides a fascinating reconstruction of the history of deer in North America during the past two thousand years. Yes, from BC to now (no, there's no mention of Fred Flintstone).

Deer history goes back much further, of course. If we recall that eighteen thousand years ago a mile-thick glacier covered much of the Midwest, we are reminded that deer habitat was displaced to the south. Deer can't gain calories from glacial ice. Their habitat was also compressed in size, meaning there must have been less habitat than there is today, and likely fewer deer. However, we know that people were harvesting deer then because of eighteen-thousand-year-old deer remains recovered from an archaeological site in Texas. Many such sites contain remains of whitetails, showing that Native Americans routinely hunted deer. This negates claims by some that Europeans were the first to hunt whitetails in North America.

Why stop at eighteen thousand years ago? There are four-million-year-old fossil deer very similar to modern whitetails. However, no clear data on populations of whitetails go back that far, so Adams and Hamilton pieced together information for the past two thousand years. As you go back in time, the information gets less and less clear, or scientifically speaking, the confidence intervals increase. Their work provides a starting point, but we need to recognize that there are quite a few guesstimates in their reconstruction of deer history. Even *NCIS* detectives lament old crime scenes where the trace grows dim.

A graph of Adams and Hamilton's data shows the estimated number of deer in the United States and Canada over the past two thousand years. The gray area indicates a range of population sizes, rightly reflecting uncertainty in deer numbers and harvest. During the period leading up to 1500, it is thought that whitetails were rather abundant and that many native tribes hunted deer year-around. They depended on deer for food, footwear, and clothing. Antlers were used as ornaments and tools, and sinews were employed as bowstrings and fishing lines. Later deer became important in Native American commerce.

One wildlife biologist estimated that a band of one hundred Native Americans would use about 950 adult deer per year, or 9.5 deer per person. I estimated

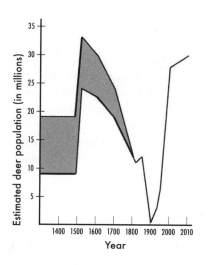

The abundance of white-tailed deer in the United States and Canada during the past two thousand years. The gray area indicates estimates of the range in populations and reflects uncertainty due to relatively poor historical information. Data from Adams and Hamilton.

that if my family of four ate nothing but venison, we could go through at least 20 deer a year, or 5 each. So 9.5 deer per Native American might not be too far off. Another author used the size of the whitetail range, the number of native people, and the amount of venison likely consumed per person to estimate that between 4.6 and 6.4 million deer were harvested by Native Americans annually. Deer biologists think that between 30 and 50 percent of the herd was harvested each year by Native Americans, and lean toward the 30 percent figure. If you know the approximate number harvested and what percentage of the total herd it represented, you can figure out how many deer were around.

For comparison, hunters in the United States and Canada took less than 7 million whitetails in 2008, or roughly 23 percent of the total number thought to exist. The human population of the United States and Canada in 2008 was 337.8 million people. Given the 2008 deer harvest, 0.02 deer were taken per person. Obviously, we no longer depend on deer for subsistence. Nor could we, because there are simply not enough deer to support our human population. To attain the annual rate of harvest of 9.5 deer per person estimated for Native Americans, the deer population today would have to be almost 14 billion, assuming a 23 percent harvest. We might not appreciate having four hundred times more deer on today's landscape.

Assuming a 30 to 50 percent harvest rate by Native Americans and given uncertainty in the estimates, the graph suggests that there were between 9 and 19 million deer in the United States and Canada up until about 1500. Native Americans were spread throughout the continent, and up to two-thirds of the continental U.S. landscape was used for some form of agriculture. The arid Southwest was terraced and irrigated. Native Americans even used prescribed burns to manage the landscape. They also deliberately used fires to push deer to waiting hunters in the autumn, as the vegetation was dry and easily burned, and the deer in the fall provided high-quality skins.

The next obvious change visible on the graph is a large increase in deer numbers beginning around 1500. Some attribute the increase to the loss of at least 20 percent of Native Americans as a result of the arrival of new diseases from Europe (thanks in part to Columbus). The reduction in human density could have allowed whitetails an "ecological release," and their populations might have doubled, reaching nearly 35 million animals. Again, however, notice on the graph that the range is from 24,000,000 to 33,000,000. It's not easy to estimate numbers of deer with much greater precision at a time that long ago. Incidentally, populations of several other kinds of animals, such as bison, elk, and the passenger pigeon, might have increased at the same time owing to the decrease in Native American populations.

Starting soon after the big increase, we see the estimated population start to fall, and the range of the estimates gets narrower. There were fewer Native Americans, but now they used deer to trade for European goods—deer had become a commodity. For nearly one hundred years beginning in the 1600s, Native Americans helped supply over 100,000 deer hides annually from the port of Charles Town, South Carolina, to leather factories on the other side of the pond. Native Americans also used deer to trade for "geegaws [trinkets], metalwares, guns, alcohol, textiles, and promises." By the end of this decline in 1800, the deer population estimate is around 12,000,000—about where the population was before it began the next significant increase.

People took notice of deer and their population trends, given how important they were to the local food base and economy. Settlers in the mid-seventeenth century noticed fewer deer. Some states enacted legislation to protect deer. Rhode Island prohibited hunting in 1646, and in 1679 New Jersey made it illegal to export deer skins from deer killed by Native Americans. Virginia outlawed harvest of does in 1738, and in 1788 New York no longer allowed hounds to be used in deer hunting. But more than two hundred years would pass before the first game wardens were hired—the first were in California and New Hampshire. Year-around deer hunting continued in most places without active enforcement.

From about 1800 to 1850, the long-term declining trend in the deer population reversed. In part, this was due to the diminished use of deer hides used for trade. Too few deer were left around settlements to make large-scale hunting profitable. Supply and demand economics might have helped the deer during this period.

The population upturn was short-lived, and the graph shows a precipitous decline of deer from 1850 to 1900. With improved means of transportation, people moved around more, and European settlers provided a great demand for meat and hides. Loss of forests and repeated burning of cutover forests resulted in fewer areas where deer could find refuge from constant hunting pressure. Lack of game laws, or not enforcing those in place, made it tough going for whitetails. By the late 1800s whitetail populations were at the lowest number in recorded history, possibly as few as 300,000 animals, or about 1 percent of their peak abundance.

Big government came to the aid of deer. One of the most important pieces of legislation for deer (and many other species) was the Lacey Act of 1900, which prevented the interstate transport of illegally killed wildlife. Now federal law supported the game laws of individual states. The act helped quell market hunting of many species, as no longer could deer, or barrels of passenger pigeons or Eskimo curlews, be shipped across state lines to large eastern markets.

In the late 1890s many states formed wildlife agencies charged with enforcing game laws, and in the 1930s University of Wisconsin's Aldo Leopold basically invented the field of scientific wildlife management. Many students were trained in the principles of wildlife management, and state agencies hired them. Also benefiting wildlife, including deer, was the formation of the National Wildlife Refuge system. As of 2009 the 550 National Wildlife Refuges protected 148,263,230 acres. The U.S. Forest Service and National Park Service were instituted, providing further means for protecting wildlife. The Pitman–Robertson Act of 1937 was of enormous value—about 10 percent of all money that sportsmen spend on hunting and fishing equipment is returned to states for wildlife restoration, based on a function of the state's total area and number of licensed hunters. These and other factors set the stage for the recovery of whitetails to a level that may at present be near the highest in recorded history (see graph).

Some of the recovery of whitetails, or at least its speed, is a result of restocking programs. Texas, for example, brought in over 16,000 deer. Most of this restocking was completed by the early 1970s, but in some states (e.g., Alabama, Kentucky, Virginia) transplants continued into the early 1990s. Restocking occurred before concern over Chronic Wasting Disease; otherwise the large-scale relocation of deer could have been a major issue. A further problem with relocations could result if populations had evolved adaptations to their local environmental conditions; relocations could break up these adaptations by introducing genes from deer adapted to conditions in other areas.

As the number of deer increased, game managers were careful to protect them. Most states and provinces instituted what is now called "Traditional Deer Management"—the basic principle of which is to prohibit harvest of does. The population grew at a rapid rate, and by the early 1970s, many herds were more abundant than the habitat could support. Deer raided gardens, and deer-vehicle collisions increased. In many areas deer almost achieved pest status, and reducing the population of does became necessary. Reversing the Traditional Deer Management mentality has not been easy. Even when requested, doe harvest in some areas reached only 20 percent of the harvest. The don't-shoot-a-doe attitude is still prevalent; 50 percent of Wisconsin hunters would not shoot a doe unless they were required to do so to get a buck permit.

Other factors contributed to the expansion of the whitetail population post-1950. Movement of people from cities to suburban or rural areas reduced available hunting land near cities, while providing food and safety for deer, and antihunting attitudes were prevalent among new residents. Some have suggested that with the extinction of the passenger pigeon, far more acorns were available for deer, which might have contributed to their population increase. This seems unlikely given that the pigeons were gone by the early twentieth century, and the increase in deer was much later, although there could be a "lag" effect. Removal of major predators such as wolves and cougars from the ecosystem also played a role in allowing population expansion. At this time, it is not clear whether the deer population will continue to increase or whether managers will achieve a stable population.

The huge swings in the historical deer population record show that deer can adapt to human-dominated landscapes. They also justify concern. If it is true that whitetails are as numerous at present as they have been throughout recorded history (see graph), we may need to be especially vigilant. The average body condition of whitetails whose abundance approaches the maximum carrying capacity of their habitat could decline; that is, there may be less for them to eat per capita, although we do not know what the carrying capacity of deer is on a continent-wide basis. Also, currently known or future diseases could be especially hard on deer at high levels of abundance. It will be interesting to see what if any consequences the current high level of abundance will have for the future of whitetails in North America. Will the next two thousand years bring more ups and downs or more stability?

2

A MESSAGE FROM OUR NATIVE BIRDS
DEER HUNTERS NEEDED

The white-tailed deer populations in many parts of its range are currently at their highest levels since the 1600s. Mild winters, lots of food, and reduced natural predation have led to deer almost reaching pest status in some areas. Many areas have liberal bag limits in an attempt to keep the herd at lower levels. But why? The point is to limit ecosystem effects, reduce potential disease spread, and reduce deer-vehicle collisions (DVC). According to an insurance company, the likelihood of a DVC in the lower forty-eight states ranges from one collision for every 1,892 vehicles (Arizona) to one in 39 (West Virginia). But what exactly are the negative eco-system effects they wish to limit?

In short, overgrazing native vegetation leads to changes that cascade throughout the food chain. We have lived for a long time in a deer-dominated landscape, and none of us have memories of what our forests looked like before European settlements. There-fore, we cannot be sure what "natural" really is, and a retrospec-tive view might be a picture of the land before Native Americans. To be safe, we would need to know what the deer herd looked like twenty thousand years ago, although much of Minnesota was then covered by a mile-thick glacier. Anyway, "natural" is relative. Nevertheless, we know that the appearance of our forests today is probably unlike it was in historical times, as a result of the food required by our expanded deer herd. It is one thing to think this, another to show it scientifically.

Documenting specific ecological effects of deer requires long-term observations or "natural" experiments. Such an experi-ment has taken place in the Queen Charlotte Islands (now called Haida Gwaii) involving black-tailed deer and songbirds. Black-tailed deer were introduced to some of the islands in the late nine-teenth century. These islands now have plants and animals that are greatly different from those on the islands without deer. Sur-veys of birds found that the abundance of songbirds was 55 to 70 percent lower on islands with the longest history of browsing than on deer-free islands. Also, predation of nests by crows, ravens, and jays was greater where concealment was reduced due to deer browsing. Clearly, birds on these islands need more deer hunters.

High deer populations are affecting the local ecology in other regions. The United Kingdom is experiencing a problem with overabundant deer. Unlike Minnesota, the United Kingdom has two native species of deer, red deer and roe deer, and four intro-duced species, fallow, muntjac, sika, and Chinese water deer (the latter ironically called "CWD," which we know as Chronic

Wasting Disease). Populations of native and feral introduced deer have been increasing rapidly and may be at a thousand-year high (always humbling to see how far back "history" goes to those in Europe). The abundance of deer in the United Kingdom has led to damage to seminatural habitats, damage to woodland plants and agricultural and horticultural crops, increased traffic accidents, and infection with diseases that can affect farm livestock and in some cases humans. All of these impacts are familiar to Minnesota residents. Although the United Kingdom has liberal deer seasons, unlike in Minnesota you cannot use a bow (it is considered "cruel"), but you can sell the meat.

The high abundance of deer has had effects on local birds, as reported in a scientific paper in the British journal *Ibis*. British ornithologists Robin Gill and Robert Fuller compared densities of songbirds in areas that had different densities of deer, ranging from deer exclosures to areas of comparable size that did not exclude deer. In the fenced plots (deer excluded), bramble cover, canopy cover, low vegetation cover, field layer density, and shrub layer density were higher. Grass cover, however, was higher outside the fences. Changes in bird numbers were equally evident. Birds that nest in low shrubs averaged 8.5 territories in the deer exclosures relative to 2.2 in the unfenced plots. This is strong evidence that overbrowsing by deer can alter the habitat enough to reduce the number of birds present.

Opening up the understory does yield habitat for other species. So the news is not all bad for birds. But in general changing the natural vegetation "unnaturally" puts the ecosystem in a new, nonnative state. Consider the fact that cities make good habitat for pigeons, and although we have a lot of them, it's not something we brag about.

None of this should be taken as a message that we necessarily have too many deer everywhere. It is clearly relative. The information on deer-bird relationships on the Queen Charlottes is from small islands, where ecological effects are often magnified. I don't think that anyone in the United States is claiming that white-tailed deer populations are endangering any bird species, although locally numbers have likely been reduced. Around Minneapolis and St. Paul, an unlimited antlerless harvest has recently been allowed owing to the high survival of deer, which is not typical of the entire state. Of course, part of the motivation to cull the metropolitan herd is ironically to permit the maintenance of nonnative (i.e., ornamental) vegetation that deer like to browse in people's yards. Also a lot of DVCs are occurring, which is generally bad, unless you own an auto body shop.

Whether hunters can effectively manage the herds in the United States, Canada, and England remains to be seen. Our habitats have been altered by a large deer herd, in some areas larger than were historically present, and while that might not be permanent, it does affect other native animals such as songbirds. Further, when the native understory is reduced or eliminated by deer, it often favors spread of introduced plants like buckthorn. Couple that with habitat loss and fragmentation, pollution and feral cats, and it just keeps getting harder to be a bird! In this regard, a deer hunter just might be a bird's best friend.

3

THE SCIENCE OF CHRONIC WASTING DISEASE AND ITS RELEVANCE FOR MANAGEMENT OF WHITE-TAILED DEER

Chronic Wasting Disease (CWD) has been found for several decades in some western states but more recently has become established in the Wisconsin white-tailed deer population, especially in the area just west of Madison. Soon after it was discovered, CWD was big news because of the high fatality rate in infected deer. Naturally, people in neighboring states are wondering whether the disease will spread, and if so, what fate might befall their white-tailed deer population. Will CWD lead to annihilation? Will it be a minor inconvenience? Will it take huge sums of money to contain and control? Can people get CWD from eating infected deer? What we know about the disease is not yet sufficient to provide definitive answers to these questions.

I think there is a lot of confusion about this important affliction of deer, and since my family and I consume large amounts of venison, I became personally interested in the subject and did a lot of digging into it.

WHAT EXACTLY IS CWD?

In a nutshell, CWD is a disease caused by a normal protein that "goes bad." It results from the abnormal folding of an otherwise normal protein into one that is called a prion (short for "proteinaceous infectious particle"). Prions, or misfolded proteins, serve no useful function and build up in the body, lodging in cells, particularly in the brain, and cause death via a major neurological meltdown.

Prion diseases in general are termed "transmissible spongiform encephalopathies," or TSEs. The cow version is "mad cow" disease. In sheep, the prion-caused disease is scrapie. In humans, one version is Creutzfeldt-Jakob disease.

Prions are not living organisms like bacteria or viruses. They do not reproduce themselves, as they have no DNA; that is, they do not divide or mate with other prions or contribute some of their material to another prion to produce a baby prion. They are in fact celibate! When a prion physically contacts a normal protein in a deer's body, it causes the normal protein to misfold. Thus, spread within an individual deer is caused by direct contact between a prion and a normal version of the protein. In this way, prions are very much unlike other disease organisms we are familiar with. Furthermore, because the disease may not reach the final stages for eighteen months or more, a doe may have time to reproduce at least once, which may not be good for the population if she is genetically predisposed to getting CWD.

HOW DO YOU KNOW IF A DEER HAS CWD?

A CWD-infected deer in the terminal stages exhibits a number of behavioral symptoms. Immediately prior to death, animals show listlessness, lowering of the head, repetitive walking in set patterns, excessive drinking and salivation, and grinding of the teeth. These behaviors are not conclusive evidence of CWD. Other diseases of deer can cause similar symptoms, and the only way at present to be sure if a deer is CWD positive (CWD+) is to autopsy the animal. Diagnosis involves a microscopic examination of the brain, tonsils, or lymph nodes performed after death. A deer that died from CWD has abnormal clumps of prions that have lodged in the brain.

What about the early stages of infection? Do deer exhibit outward signs of CWD, or can prions be detected in minute levels in blood or lymph fluid? In theory it should be possible to detect minute levels of prions in feces, urine, blood, or tissue—we can isolate DNA from a single hair. Some promising research results are on the horizon that try to detect prions in low concentrations in blood or feces. At present there are no easy tests. For example, a recent scientific study showed that deer excrete prions in their feces before they show the outward behavioral symptoms of late-stage CWD. However, it is not a simple matter of examining a fecal sample. Instead, the feces are homogenized into a solution and injected into the brains of transgenic mice, and then the mice are sacrificed months later and their brains examined for clumps of prions. Owners of game farms will especially welcome when definitive tests can be done on asymptomatic deer so that they have an early detection system for their captive herds. Detecting prions in tissue or blood might also eventually be cheaper than sectioning brains—the University of Wisconsin at Madison presently charges $50 "a head" (literally) to test deer.

We know that a deer may not show the outward symptoms of CWD for several years, and while it is "asymptomatic," it can shed prions, much like how we can shed flu virus via sneezing before we feel sick. Unfortunately, at present we cannot tell a deer has CWD until the final stages.

HOW DO DEER GET CWD?

An animal can get infectious prions from direct contact with another animal and from prions in the soil via urine or feces, or it might develop them spontaneously. I asked an expert on prions, Charles Weissmann, head of the Scripps Florida Department of Infectology, about spontaneous development, and he said, "There is, to my knowledge, no data on deer. For sure cattle and humans can develop prion disease spontaneously; the frequency for humans is about one in a million." So if that frequency were also true for deer, then a few of the several million in the upper Midwest would get the disease without contacting an infected animal.

An article in the journal *Science* in 2006 by Candace K. Mathiason and sixteen coauthors reported that uninfected white-tailed fawns could be infected with CWD by giving them oral doses of saliva from infected deer. Now, few fawns were tested (ten), and the dose of saliva containing a high concentration of infectious prions, two-thirds cup, is more than an individual deer is likely to encounter under natural circumstances.

The Mathiason group published a follow-up study in 2009 in *PLOS ONE*. They again gave naive fawns saliva from infected deer, and two-thirds got it. They commented, however, that their methods involved "likely unrealistic doses [of saliva] to be acquired in a natural setting." But these studies showed that deer could get CWD from the saliva of an infected deer.

These studies by Mathiason and her colleagues are a main source of evidence that deer might transfer CWD by touching noses or eating from a contaminated food source, and thus they provide the scientific basis for bans on feeding and baiting to prevent transmission of CWD. Still we do not know how much saliva is required for transmission, or at what stage in the infection a deer's saliva becomes infectious. Of course, at this time the "safe" answer is "it doesn't matter"; just assume that any contact from any deer at any stage of infection is risky. More recently, a study showed that a dose twenty times less concentrated than the oral doses administered could be inhaled by a healthy deer and result in an infection. Even more recently, a study showed that growing plants can incorporate prions into their tissues, which can then be eaten by deer! Although Mathiason's group failed to show that fawns contracted CWD from urine or feces from infected deer, a different study showed that asymptomatic deer excrete infectious prions in feces. This study was important because it established one way in which prions could build up in the soil, where uninfected deer could pick them up. However, when during the course of infection deer shed enough prions via feces to infect wild deer is unclear, as is how long prions remain in the soil (at least a year) and whether deer always get CWD from prions in the soil if they are exposed. That is, what is a "fatal dose"?

Deer might transmit prions by using the same bedding spots, but no scientific studies have investigated this phenomenon in the wild. The best information is from the second study by Mathiason's group, where they inoculated bedding with CWD prions, exposed two naive deer to the bedding, and found that CWD developed in fifteen months in one, and in nineteen months in the second. The two deer used the same bedding for 570 straight days, and the bedding was "refreshed" daily with prions. These are not natural conditions, but the study establishes the potential for environmental contamination and transfer to healthy deer.

We don't know whether a single infected deer that wanders over a large area will concentrate enough prions to make it likely that other wild deer will pick them up. Also we don't know how far along the disease has to be before prions in the feces are sufficiently concentrated. Safe to assume that any are bad, but there's lots to do here too.

PEOPLE AND CWD

An ongoing concern is whether humans who eat CWD-infected deer could get a similar disorder. The question here is whether the disease can cross the "species barrier." There is no known instance of a person getting CWD from eating infected deer or elk. Deer and human prions are not closely related (i.e., those that cause Creutzfeldt-Jakob disease in humans differ from those causing CWD), which reduces risk to people.

Humans can get mad cow disease, however, so there remains concern about CWD. Cows and deer often share pastures, further elevating concern. In a study published in *Veterinary Pathology* in 2007, A. N. Hamir tested whether cattle could get CWD. They drilled a hole in the skull and injected a concentrated solution of prions into the brain, so predicting from this study whether cows can get the disease under normal circumstances would be a stretch. The cows were sickened but did not show the same brain disintegration typical of the disease. Probably drilling the hole in the skull didn't help.

Experiments with human cells in a petri dish (the brain inoculation route is out!) show that a "molecular barrier" appears to ward off CWD prions. That's good news. To be on the safe side, most authorities recommend not eating lymph nodes, brain, and spinal cord of deer, as that is where the prions are concentrated.

Ideally, though, from a scientific perspective, we would need to follow people through time who had consumed CWD+ venison. Of course, scientists could not do this knowingly, because of ethical and legal reasons. And they are not likely to get many (any?) volunteers.

However, on March 13, 2005, a local fire company in upstate New York hosted a fund-raiser at which venison was served. It was later learned that one of the deer had CWD. New York has a law that says all venison served at such events must be tested, but the test doesn't have to precede the event! A group of health scientists is now monitoring eighty-one people who either ate the venison or were in some way exposed, and some results should be available soon. This incident should be a potent lesson to any organization offering venison or elk at a banquet. The Centers for Disease Control and Prevention recommends not eating meat from an infected deer.

CWD AS A BACTERIAL DISEASE?

The newest player in the CWD game is a cell wall–less bacterium called a spiroplasma that is usually associated with insects and plants. These spiroplasmas seem to cause the same brain deterioration that rodents show when experimentally infected with scrapie (the sheep version of CWD).

A study published in the journal *Veterinary Ophthalmology* in 2011 by Frank Bastian from Louisiana State University (Baton Rouge) suggested that the cause of CWD was this spiroplasma. If that is true, we've been barking up the totally wrong tree. However, Bastian has been advocating this view for two decades, and his work remains controversial; other labs have been unable to replicate his results. My question is, how would the bacterium survive in the soil? But what if he's right? We should not close the door on any ideas at this time, as the science, albeit suggestive, is too uncertain to support definitive conclusions about transmission.

ARE GAME FARMS CANARIES IN THE COAL MINE?

Minnesota's only wild CWD+ deer was harvested in the area where a former elk farm experienced an outbreak of the disease a few years prior. This has accentuated the call from many people for greater controls, if not outright elimination, of game farms. I think that this sentiment is misguided. CWD is a disease of wild cervids. It does not appear in farmed deer and elk just because they are captive. Many people against game farms argue as though the game farms are to blame for the disease. This is not the case. Deer do not get CWD simply because they are in captivity.

Two questions are important when considering CWD. The first is how it originates, which we dealt with above. The second question is how it is spread around the landscape. Many believe that the sale and transport of captive deer result in the spread of CWD owing to the fact that unless a deer is in the late stages of infection, one would not know it was infected. In the wild, high population densities clearly facilitate its transmission. Moose, for example, are considered relatively safe from CWD because they are so spread out, probably like deer once were.

In captive herds CWD stands out. First, because infected animals shed prions before they show outward symptoms, prions can reach very high, perhaps abnormally so, concentrations in the soil of game farms, making it easier for other animals in the captive herd to become infected in their relatively confined space.

Most important, captive animals often live longer lives than wild deer. CWD takes up to several years to reach the last stages. Deer in captive herds reach older ages and are therefore more likely to die from the infection. Wild deer, on the other hand, may not live long enough to show the final stages of CWD because they are killed by a predator or a hunter or they die in the woods and are not found. It is known that mountain lions preferentially prey on CWD+ mule deer. Not many wild deer live to be over four to five years. So we are more likely to observe the disease in captive herds.

Perhaps captive and wild animals come into contact along fences, because wild deer that are sick might be drawn to captive facilities where it appears that life might be easy, or at least food is readily available. A wild animal with CWD could transfer the disease to a captive deer and go off and die or be harvested. Later, we see the disease in the captive herd because the infected animals are confined to the game farm and live long enough to show symptoms.

Thus, when a captive herd develops CWD, it may be an indication that the disease is in the local wild population—hence the captive herd is the canary in the mine. Checking the sources of the infected captive animals to ensure the disease wasn't transferred from another facility is vital. But the surrounding wild herd should also be checked, as it could be the source. After all, it would be more than ironic if the state were sued by a game farm operator whose captive deer were infected with CWD by a wild deer.

I do not preach that game farmers should lose their right to a livelihood. I do think that they should have strong precautions in place to prevent the escape of captive deer. Most important, research aimed at detecting deer in the early stages of infection is needed to prevent transport of early-stage infected animals from farm to farm and state to state.

WHAT WILL THE CONSEQUENCES BE FOR THE POPULATION?

The disease has been around long enough in some areas that some statistics have emerged. For example, CWD was found in whitetails in Saskatchewan in 2002, and through 2008 the rate of infection stayed at or less than 1.5 percent. Data from Colorado are available for 2006 through 2008 online. Of 308 deer tested, 8 were positive (2.6 percent). The biggest sample comes from Wisconsin, where 166,472 wild deer were tested statewide from 1999 through 2011; 1,564 were positive, less than 1 percent (0.9 percent). In the CWD core area, just west of Madison, the frequency is 1.3 percent (1,568 positive in 118,095). These levels do not seem too high, given the concern about CWD.

These figures are misleading, however, if taken out of context. A closer look at the Wisconsin data (see http://dnr.wi.gov/topic/wildlifehabitat/results.html) is instructive, as the infection statistics can be viewed in a number of ways. The core area of concern includes eastern Iowa and western Dane Counties, where you can see the data by square mile sections. Although CWD was detected relatively recently, some sections have infection rates as high as 23 percent. Now that is obviously a grave concern, and the Wisconsin Department of Natural Resources is taking it seriously.

Still, the percentage of infected deer varies depending on the area and year. In the core area, for example, infection frequency is at 12 percent in bucks, and 6.5 percent in does. A colleague and I have spent hours debating why the rate is higher in bucks than does, but we haven't arrived at any firm conclusions (maybe it's as simple as the bucks are just generally older). In the persistently affected deer management area 70A, CWD frequency has continued to increase.

The data contain many interesting features. For instance, the frequency of yearling does with CWD is approaching that of adult females, which is disturbing because scientists have assumed that young deer don't get CWD and don't test them. A recent genetic study found that related does tend to have higher infection rates, suggesting transmission through animal-to-animal contact rather than through the environment. Because related does hang around each other, attaining a representative sample of the population for testing can be difficult. The relatively high percentage of males with CWD coupled with their larger home ranges has prompted some to suggest selectively culling young males.

Diseases have an easier time being diseases when their host populations are relatively common. The number of white-tailed deer in North America is likely at an all-time record high. For those of you who like this trend and would favor even further growth, diseases of deer should command your attention. A 2011 scientific paper by Emily Almberg and colleagues in the journal *PLOS ONE* examined various models that predict the outcome of CWD infection. For any model, many "parameters," or inputs, need to be considered. For CWD, they include the typical things like deer numbers, movement patterns, reproductive output, life span, and sex and age ratios. The additional concern for CWD is that the infectious agent, a prion, can remain in the soil for long periods and serve as a long-term reservoir that can infect healthy deer. In other words, a prion doesn't die

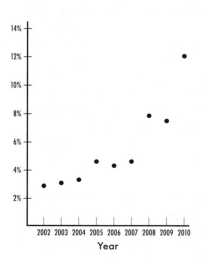

Frequency of CWD in deer in Wisconsin's zone 70A. The average for these years is 4.5 percent, but recently the proportion of deer infected has climbed.

with its "host" but can outlive it for some time in the environment.

Almberg and her group concluded the following: "Resulting long-term outcomes range from relatively low disease prevalence and limited host-population decline to host-population collapse and extinction." Obviously these are vastly different outcomes. In one case we don't need to worry, but in the other we'll all be hunting rabbits, raccoons, and squirrels.

CAN A DEER'S GENETICS SAVE IT?

We are all familiar with the potential of insect pests to evolve resistance to pesticides. You get a new flu shot each year for the same reason—the virus evolves in response to vaccines. People are gaining new familiarity with bed bugs owing in part to the insect's evolved resistance to DDT (which we've quit using) and pyrethroids. A few bed bugs gained a genetic mutation that altered a natural protein in their body in a way that allows it to detoxify pyrethroids. This allows bed bugs to be unaffected (immune in a sense) by these chemical pesticides, and their offspring to proliferate. "Sleep tight, don't let the bed bugs bite" is becoming more apt advice.

Could deer have some form of natural immunity? Researchers found several different versions of the genetic instructions encoded in the DNA of deer for making the normal protein. These versions are called alleles, and each animal (including us) carries two copies in each cell. This means that an animal can have various genetic combinations (termed genotypes). If there are three alleles (A, B, C) the genotypes could be AA, AB, AC, BB, BC, or CC. In Wisconsin, one relatively rare genotype was underrepresented in the CWD+ deer. Although at least a few deer had CWD no matter what their genotype, a couple of genotypes apparently show a degree of resistance to CWD. Thus, as in bed bugs, some deer may be genetically resistant to the disease, and one might expect deer with the resistant genotype to increase over time.

I was interested in the prion genetics of Minnesota deer. I

gathered fifteen tissue samples from deer taken in the Twin Cities metropolitan area in the past couple of years, and my lab at the University of Minnesota (thanks to Mike Westberg) sequenced the prion gene. We found that two of the deer possessed the genotype determined from studies in Wisconsin to confer some degree of resistance, for a frequency of 13 percent. This does not mean that 13 percent are CWD resistant, but it provides some hope that deer genetics might come into play. However, if deer reproduce before dying of CWD, then those genotypes will not increase in the population. In the future we hope to expand this work statewide and produce a contour map of CWD-resistant genotypes, as well as use genetic markers to estimate whether deer use some routes of dispersal more than others. The latter information would provide a means for dealing with any potential outbreaks by showing where to target monitoring or culling.

WHERE ARE WE AT PRESENT AND WHAT DON'T WE KNOW ABOUT CWD?

No one doubts that CWD has spread in Wisconsin and has the potential to be very bad for deer population numbers. We should be greatly concerned that at a local scale infection rates have reached 20 percent in Wisconsin and could go higher. Maybe such high frequencies will be very local phenomena. Unfortunately a natural experiment is under way in Wisconsin, and we need to watch closely how the disease progresses. We will also be watching the monitoring of the people in New York who became unknowing participants in a study of the transmission of CWD to humans. Lots of new research is on the horizon, so we can hope for efficient ways to monitor and deal with the disease in wild herds. Let's hope the spread is contained.

In sum, several important questions about CWD remain unanswered. How many prions does a deer have to ingest to become infected? How long after infection can a deer spread enough prions to infect other deer? Is the main avenue of spreading via saliva or soil or plants? These together with a lack of an early detection system for the disease in captive deer make CWD a potentially critical issue throughout the range of deer in North America.

4

URBAN DEER
HUNTING VERSUS BIRTH CONTROL

Many urban deer herds are at all-time high population levels. In many areas, wildlife agencies issue unlimited permits for antlerless deer. From a biological perspective, when there are too many mouths to feed, less food goes in each, so individual birth rates fall, and death rates rise. When these factors balance, individuals on average leave one offspring, and in the jargon of ecologists, we say that the population is at its carrying capacity and is relatively stable. In the suburbs, we tend to plant stuff that deer like to eat, and we remove big predators. These actions tend to increase deer abundance. Human reactions to this new higher equilibrium number of deer depend on whether people enjoy seeing deer or just had the front end of a new vehicle replaced.

In the Twin Cities metropolitan area and others, the question arises as to how to control high deer numbers. One method is a liberal hunting opportunity. Some urban parks that are normally closed to hunting conduct permitted bow hunts or hire sharpshooters to kill deer. The former option provides opportunities for many hunters that might not otherwise be able to hunt. However, the antihunting public sometimes reacts negatively to hunting virtually in their backyards. How to deal with large numbers of deer is becoming an important issue as suburbs expand and encroach on more and more land, putting humans and deer in greater and greater conflict.

The use of contraceptives is another method to reduce deer.

The public often favors this means of herd reduction because it reduces birth rates and is viewed as more "humane." I, like probably a lot of others, responded to questions about whether we can manage high-density deer herds with contraceptives by saying it doesn't work. However, I have to admit that I gave this response based on hearsay. So, when a scientific review of contraceptive use in wild animals appeared, I took a crash course. What follows is a summary of a paper in the journal *Integrative Zoology* by Kathleen Fagerstone of the U.S. Department of Agriculture and colleagues.

These authors review the ways in which contraceptives can be used to limit deer numbers. Two main methods are to administer a chemical contraceptive via food (orally) or directly via injections or implants. Each has its advantages and disadvantages. Oral contraceptives are easy to distribute but generally considered ineffective or impractical for deer because the deer must consume them regularly, plus other organisms might ingest them with unintended detrimental effects.

Several contraceptives have been injected. One category includes steroid hormones, which have been shown to reduce or inhibit estrous behavior in does. Although this method inhibits reproduction in whitetails (by interfering with ovulation or implantation in does and by impairing sperm production in bucks), it requires continuous doses from an implant or frequent injections. Thus, because their effects are short-term, steroids are not currently considered a viable contraceptive.

A second class of chemical contraceptives is "immunocontraception vaccines." These cause an animal to produce antibodies against their own reproductive proteins and can last from one to four years. The vaccine SpayVac, for example, has been effective in preventing reproduction in does for up to four years after a single shot, an advantage over many others that require annual booster shots (and like many people, deer lack health insurance). Although it prevents pregnancy, SpayVac can make does have multiple or back-to-back estrous cycles and thereby extend the potential "breeding" season. A different study evaluated the possibility that this would cause increased deer-vehicle collisions, because bucks would be chasing does across highways for an extended rut period, but the researchers found no evidence for it. SpayVac is not currently used to manage whitetails in the United States.

Another vaccine is GonaCon Immunocontraceptive Vaccine. A single injection reduces or eliminates reproduction in both sexes of white-tailed deer. In bucks, the vaccine reduces testosterone levels, testicular size, and aggressive behavior, which results in "no interest" in estrous females. Bucks treated with a different vaccine either dropped their antlers early or remained in velvet.

Fagerstone and colleagues ascertained that these vaccines caused no long-term health problems in treated deer, and apparently these vaccines do not harm people who eat treated deer. I was concerned, however, that some vaccines "reduced progesterone concentrations, altered estrus behavior, contraception, failure to maintain pregnancy following conception, and reduced fawning rates." Now, maybe as a hunter (and bow hunter at that) I'm not allowed to offer an opinion about what's humane and what's not. But administering a contraceptive that causes these effects, including spontaneous abortion, does not strike me as terribly natural or humane, and these consequences should be made clear to those favoring this method of herd control.

There are legal aspects to this issue. Contraceptives for all wildlife are regulated at present by the U.S. Environmental Protection Agency (EPA). Getting approval is not trivial and can take up to three years. At present, the EPA has approved only two contraceptives for wildlife (believe it or not, they are actually called "reproductive pesticides"). One is an oral contraceptive for managing Canada geese and pigeons, called OvoControl G. The other is the GonaCon Immunocontraceptive Vaccine, mentioned above, which was approved in September 2009. These are both considered restricted-use products and can be used only by persons that have been certified in their delivery. These restrictions are to ensure humane capture and treatment of animals, awareness of potential hazards to the person administering the vaccine (i.e., what happens to you if you fall on the needle), and prevention of misuse on nontarget species (i.e., that annoying neighbor dog . . .).

What are some of the reasons that these contraceptives are not in wide use? Fagerstone and her colleagues discussed the biological and economic feasibility of contraceptives. First, statistical wildlife models show that reproductive control (as opposed to lethal control) will be most effective for species with high reproductive rates and low survival. These include animals like rodents and some birds. The authors wrote that "reproductive control will typically be less efficient than lethal control [hunting] in managing populations for larger species such as deer, coyotes and Canada geese that do not typically reproduce until 2–4 years of age" and "have smaller litter or clutch sizes than most rodents." In practice, more than 50 percent of fertile does need to be kept infertile to reduce population numbers. If the vaccine must be administered yearly, 60 to 80 percent of all does would require vaccination. Some chemical contraceptives require continuous doses and necessitate either annual darting or capture and implantation of a sustained-release capsule. A second, related drawback to large-scale deer contraceptive efforts is the cost, which can be around $250 for treating a single deer. Local municipalities would have to pay a substantial amount for treatment, a cost that would be recurring. Even if treated does are infertile for up to four years postvaccination, immigration from neighboring areas will add new, unvaccinated does to the local population.

Fagerstone and her colleagues wrote that "many wildlife agencies and biologists have been reluctant to acknowledge the potential applicability of fertility control for managing wildlife populations in part because contraceptives have been publicized as replacements for sport hunting." They point out that only 9 percent of state wildlife agencies have an established policy on wildlife contraception compared to 39 percent of fifty-four "environmental and activist groups." They concluded that "traditional methods of population reduction will still need to be applied because the cost and difficulty of delivery of contraceptive techniques would prelude their use." In a different paper, one of the Fagerstone's colleagues, Lowell Miller, who helped develop Gona-Con, expressed considerable enthusiasm, saying that the contra-ceptive was "an exciting new deer management tool," although later in the article Miller and his colleagues said that "it must be emphasized that wildlife contraception using GonaCon™ vac-cine and other infertility agents will not replace sport hunting as a wildlife population management tool. Wildlife contraception will be applied only in special situations where traditional manage-ment methods cannot be used." I'm less optimistic and think there could be pressure to use contraceptives instead of sport hunting to reduce deer herds in areas where hunting is still feasible.

The bottom line is that contraceptives do work on deer, espe-cially does, and a lot of research on their use has been and is still being done. However, the cost per deer, which includes the vac-cine and the need for trained people to administer proper doses, is economically prohibitive. Having to recapture deer and the influx of new deer from neighboring areas also render this alter-native nonviable. I also question whether it is truly ethically supe-rior to sterilize deer, given the effects on behavior and physiology noted above. The fact that contraceptives are not an economically viable way to reduce deer numbers should be useful information to communities that have outlawed bow hunting but have deer overpopulation problems. Unless they are willing to pay a large, ongoing cost, contraceptives will be an expensive way to manage a deer herd, especially when hunters will often do it for free (or at a cost to themselves of a license).

5

IT'S TAKEN CENTURIES, BUT WE NOW KNOW WHY DEER DON'T ASK TO USE YOUR COMPASS

Natural history is about learning basic facts about plants and animals, where they occur, what they do, how they interact, and so on. Many scientific journals are devoted to reporting observations that scientists make about natural history. The public, however, sometimes does not fully understand what "natural history" entails—as the editor of one prominent journal reported after receiving a letter requesting a list of nude beaches in a Latin America country.

At one time I would have wagered that we have missed relatively few "obvious" things that occur in nature. We know about migration, hibernation, what habitat birds use, who eats whom, who's the fastest, and so on. But not all things can be observed with the unaided eye, and some things that animals do require sophisticated technology for us to see or understand them.

For example, in the past twenty years we have learned that many birds have a well-developed ability to see in the ultraviolet (UV) part of the light spectrum. And scientific experiments have now well established that birds communicate visually with patches of UV plumage, which we simply cannot see. We miss much of what birds are telling each other because we are blind to the signals.

Still, you would think that something basic, something that can be observed with the unaided eye, would not have escaped centuries of observations. Right? Apparently we have missed something. It was reported by Sabine Begall (from Essen, Germany) and colleagues Jaroslav Červený, Julia Neef, Oldřich Vojtěch, and Hynek Burda in one of the most prestigious scientific journals, the *Proceedings of the National Academy of Sciences*.

But before I give it away, here's some background.

Ranchers have known for a long time that most sheep and cattle in a group face the same way when grazing. Farmers know that cattle usually face into the wind, whereas sheep face away from it. When it is cold and the sun is out, you'll probably see grazing animals orient perpendicular to the sun to absorb the most solar radiation. On a cold, windy winter's day, you'll most likely see cows (and hence the whole herd) orient themselves into a strong wind, exposing the least amount of body surface area.

These behaviors occur in stressful conditions. But what happens when conditions are not particularly stressful? One would think that the direction of body orientation would be random. Yet we now know that cows and two kinds of deer in Europe (red and roe) orient themselves very strongly in a north–south direction in nonstressful conditions! No way, how could we have missed this?

How Begall and her colleagues figured this out is pretty cool to say the least. Believe it or not, they said they looked at maps on Google Earth and that they could see cows in pastures and determine which direction they were facing. I was skeptical, so I asked Begall to send me coordinates so I could see for myself—they weren't kidding!

Try it yourself by visiting Google Earth and entering 36°58'57.57" S, 61°42'11.55" W. Sure enough you can see a herd of cows in an Argentine pasture oriented north–south, from space! They actually used images like this to figure out which way the cattle were standing.

Begall and colleagues reasoned that because their images were from Africa, Asia, Australia, Europe, North and South America, and Australia, it was unlikely that the wind and sun were the same in all places and that these could not explain the consistent north–south body orientation. They concluded that cattle were using the earth's magnetic field. We do know that lots of animals, like birds, use magnetic lines of force to navigate. Now we know that cattle and deer also can sense magnetic lines of force and use them to orient themselves.

But wait, you say, what about deer? Well, they have field

observations (not satellite) of roe deer and red deer in the Czech Republic. They found that both resting and grazing deer tend strongly to orient in a north–south direction.

Even more impressive, they examined deer beds in the snow, finding again a very strong north–south orientation! They also concluded that resting roe deer not only oriented themselves north–south but also almost always oriented their heads to point north. Somewhat comforting to the natural historian in me was the observation that in herds of grazing red deer, about one-third of the animals were oriented south, which they figured was an antipredator behavior. So, it's not all or none.

Their observations certainly raise some questions. I would think that resting deer would orient into the wind to detect approaching predators. However, the authors concluded that at least at night, bedded roe deer and red deer are deep in forests, where wind is damped and hearing becomes the main defense.

The obvious overarching question is, why? The authors also recognized this but could only make some suggestions. Perhaps maintaining a certain magnetic orientation provides a directional reference for the animals in case they are disturbed, the herd scatters, and they need to know which way to go to regroup as quickly as possible.

The authors point out that the phenomenon requires more study, but one cannot help being awed by their ending comment: "It is amazing that this ubiquitous conspicuous phenomenon apparently has remained unnoticed by herdsmen and hunters for thousands of years."

Ouch, that hurts the natural historian in me. But when we cease being amazed by nature, I guess it would stop being fun. So, now when we see whitetails feeding in nonstressful conditions and we find beds in the winter, we should record whether they are indeed oriented in a north–south direction. Remember, it's not likely all or none, but it would be interesting to know if North American deer also show a strong tendency to orient north–south. But there's a potential catch. The researchers found that the direction of orientation followed the magnetic north, which can be different from simple geographic north (in Minnesota, we're a few degrees off).

6

WHY ARE MEDICAL RESEARCHERS INTERESTED IN ANTLERS?

During a recent autumn, I occasionally saw a year-and-a-half-old buck with Y-shaped antlers (a four pointer) on each side as he walked by my trail camera. The last time I saw him he had shed one side. I found the other side while I was taking a break from writing this very article. It had been dropped in the previous twelve hours.

Wondering what is known about antlers, I did some digging in the antler literature and found that the medical profession admires them for entirely different reasons than hunters do. Limb regeneration is their interest. Yes, antlers are a lot like limbs.

Among the vertebrates, creatures like salamanders retain the ability to regenerate tails and limbs throughout life. That is a useful trick if some predator grabs your tail—better to give it up and live on with a newly regenerated one. Humans actually have some limited capability to regenerate tips of fingers, but it depends on how the wound is treated after losing a terminal segment. But the power to regenerate is pretty limited in other vertebrates, especially mammals.

With one big exception, antlers. The annual regrowth of antlers proves that mammals are not incapable of limb regeneration, but it doesn't occur, except for antlers. In a 2010 scientific paper in the journal *Gerontology*, Uwe Kierdorf and Horst Kierdorf wrote, "Understanding the mechanisms controlling antler regeneration may thus assist regenerative medicine to achieve its ultimate, but still distant, goal of inducing limb regeneration in humans." It hadn't occurred to me that limb regeneration and antlers had anything in common.

Although antlers are grown and shed annually, they actually grow from permanent features on the skull called pedicles (from the Latin meaning "little foot"). After the pedicle is formed, first antler growth starts and is associated with decreasing testosterone levels. Those who have harvested young bucks will have noticed that the first antlers are usually simple, unbranched spikes that grow as extensions of the pedicles and lack the elaborate base (coronet or burr) that mature antlers have. Typically the pedicle gets wider (and shorter) at the base with age, and bucks grow progressively larger and larger antlers until about age six, when they start to ebb.

Interestingly, the antlers begin to grow when the males are not in reproductive condition, and it is thought that increasing day length begins this process. We know that day length is a major factor because if you put deer in captivity and experimentally "speed up" the cycle of day length, red deer can grow two sets of antlers in a single year. Or if you reverse their seasons in captivity by changing their light-dark cycles by six months, you can get them to grow antlers six months out of phase.

During antler growth, the skin over the pedicle area grows outwards and becomes "velvet," which supplies blood to the developing bone. Antler growth can be very rapid—up to an inch a day in elk. Incidentally, antlers grow from the tips, not the base (like horns). This was figured out by putting one screw near the base and another near the tip of a growing antler and observing later that the one near the base was the same distance away from the base, but the screw at the tip was much farther away. Increased testosterone levels in summer stop antler growth and cause final development of the antler bone and shedding of the velvet (which the bucks help along by rubbing). Our buck is now armed and potentially dangerous or attractive, depending on whether the observer is another male or a doe (or you).

The antler itself is formed from two different types of bone, a hard outer layer (or "shaft") called cortical bone (hard, compact bone with few spaces and a porosity of 5 percent to 30 percent), and trabecular bone, sometimes called "spongy" bone owing to its porosity of up to 90 percent, which occurs in the bone's center.

The pedicle is an interesting structure, downright magical if you crave big antlers. A layer called the antlerogenic periosteum (roughly translated, antler-producing bone cover, or AP) lies over the pedicle, and its cells harbor the ability to produce the two types of antler bone. We know it has this potential because if you surgically transplant the AP to another area on the deer's body, you can induce antlers (not full ones, but antlers nonetheless) on, say, the leg! Furthermore, these "ectopic" antlers are covered with velvet, showing that the cells in the AP can transform normal skin into velvet via the chemical messages they send out. The AP cells even include the information for the direction of the axis of antler growth (so if you transplant it backwards, the antlers will grow in reverse direction).

There is some debate as to whether hard antlers are living or dead structures. Some think that blood is supplied through the interior of the antler long after the velvet is gone. They believe that this interior blood supply is cut off only weeks before antler shedding and that the two are related. The jury is still out.

After the rut, testosterone levels decrease, and the antlers are dropped. Different bucks reach this lowered testosterone threshold at different times. Just prior to shedding, a zone forms at the top of the pedicle (close inspection would show a swelling of the skin at the base of the antler) where the bone is broken down by osteoclast (literally "bone breaker") cells. This destruction of bone in this narrow zone leads to a weakening of the burr-pedicle junction, and the antler loosens and falls off.

Shedding is a relatively rapid process, and soon-to-be-shed antlers don't wobble like our teeth do before we lose them. I saw a hunting episode on TV recently where a hunter shot a large buck and sat down to prop it up for the final photo. When he grabbed the large, ten-point rack and turned the buck's head toward the camera, the antlers both came off in his hands. The hunter was not thrilled to have illustrated that when antlers are ready to come off, they do so quickly.

Now a bit of history. How do we know that dropping testosterone levels influence antler shedding. None other than Aristotle said around 350 BC that "if stags be mutilated, when, by reason of their age, they have as yet no horns, they never grow horns at all; if they be mutilated when they have horns, the horns remain unchanged in size and the animal does not lose them." Of course, *mutilation* refers to damage to the testicles. I'm often amazed at what the old guys knew.

I started this essay discussing the medical profession's interest in antlers and their role as models in limb regeneration (imagine the benefits in war times). One reason for the interest is the absence of scar tissue after the antlers are shed. Some scabbing over occurs but little else. Scar tissue would inhibit subsequent growth of the next set of antlers. If we could figure out how to eliminate scar tissue in humans, we might be able to get stem cells to "reset" to an earlier stage and begin producing bone again, possibly regenerating a lost bone. I skipped explaining a lot of cell biology here, but lack of scar tissue is one reason that antler growth, shedding, and regrowth are of interest to medical research.

A second reason is that the speed of tissue growth (via cell division) involved in producing antlers rivals that of the speed of cancer cell growth. The question is, why don't deer antlers have way more malignant tumors given how rapidly the cells proliferate (a hallmark of cancer)?

Some interesting questions remain. Why does a buck shed antlers in the first place? Growing them takes a terrific investment, and if growth could simply be suspended for the "off-season," then they could continue to grow next season. Apparently this alternative runs into some insurmountable physiological barriers, and maybe ecological and behavioral ones as well. To grow new bone, velvet and its blood supply are needed and cannot be easily reestablished on a "dead" antler. Maybe peace in the herd is not easily attained if antlers are always visible. Nonetheless, antler regeneration offers some intriguing glimpses into our own biology that may someday have major medical ramifications.

7

ISN'T IT OBVIOUS WHY DEER HAVE ANTLERS?

Antlers are bony structures, distinct from horns, unique to deer, and carried by the males of all species except the Chinese water deer and the musk deer, and by both male and female caribou.

I wager that most people have a decent idea of the function of antlers on deer: fighting. However, there have been some interesting, if not bizarre, alternative suggestions. In 1937, the German zoologist Han Krieg suggested that deer grew antlers to remove excessive minerals consumed in their diet. Others figured that excreting excess minerals in feces or urine was more likely.

In the prestigious journal *Nature*, Bernard Stonehouse noted in 1968 that growing antlers have a large number of blood vessels, no fat under the velvet, and a large surface area. He suggested that "thermoregulation may thus be the function which primarily determines the form and proportions of antlers, and necessitates their annual renewal." In other words, he thought that antlers might help deer regulate their body temperature, like ears do on an elephant. Of course if that's true, why don't females have them (they do in caribou)? Stonehouse argued that males are usually larger than females and have greater need to cool off in the heat of summer. To put it another way, a male needs to have a big body to be the boss buck, but it carries a summertime "overheating penalty," which is circumvented by growing large antlers with velvet that dissipate heat. This idea wasn't a big hit among deer biologists.

Even Charles Darwin took a stab at suggesting a function of antlers: predator protection. But others later noted that does and young animals are most vulnerable to predators, and they don't have antlers. Female caribou, however, might use their antlers to help protect their calves.

Some authors have suggested that antlers give bucks better access to food. Caribou use their antlers to expose lichens under the snow, and deer have been seen using their antlers to knock fruit from trees. These uses of antlers are probably secondary, however, because large, elaborate antlers would not be necessary to accomplish these tasks.

There was even a study that showed that the antlers of moose serve to redirect sound toward the ears, giving the male moose an added advantage of an external hearing aid!

Antlers likely have at least one function. Useless features, especially expensive ones, don't last long in nature. So having antlers must make the bearers more "fit" and give them an advantage that can be passed on to their offspring. We know, for example, that birdsong helps males to attract mates and to repel other males. Antlers aren't much different.

Use in competition for mates is probably the most obvious function of antlers, and like birdsong, antlers probably function in at least two ways. In particular, they help males to establish who is most dominant, and females prefer the most dominant bucks. It is not straightforward, however. If the size of the antlers was a perfect predictor of a buck's fighting ability and if bucks could somehow assess whether their rack was smaller than another buck's, there would be no need for fights. If a buck's rack was smaller, he would give way to the bigger buck because he would know he would lose the fight and, if injured, be out of action and forfeit any possible chance for a finding mates. Bucks have reason to be around during the rut even if they are not the biggest. A dominant buck may spend twenty-four hours with a doe in heat; during that time less dominant males often take advantage of mating opportunities.

Evidence in support of the importance of antlers in fighting comes from experiments in which antlers were removed. Typically, antlerless males fall to the bottom of the dominance hierarchy. In red deer, postbreeding bachelor groups go through a period where the last ones with antlers remaining are most dominant, but once they have all dropped their antlers, the "pecking order" is reestablished based on age and size. Alternatively, tule elk bulls in California maintain fighting ability and hold large harems, even with severe antler breakage.

Why do bucks fight then? There tends to be a strong relationship between antler and body size, such that large-bodied deer with large antlers are most dominant and hence mate with more does. But the relationship is not perfect. We know that body size is also important. Apparently as long as a buck's rack is big enough to engage another buck, the encounter becomes a shoving/wrestling match, and being stronger is likely as important as having a few more inches of antler. Despite how neat it would be, it is not possible to score a buck's antlers and then automatically know his dominance rank.

As for many opinions, scientific data on the relationship between body size, antler score, and fighting ability are lacking. Dómhnall Jennings and colleagues conducted a detailed analysis of fighting between fallow deer. One can break the question down into several components, as we have already started to. First, given a lot of first-year bucks and progressively fewer in each older age class, is there a random probability of two bucks fighting? Clearly not, as a spike (a buck with just one point on each side) is not going to challenge a five-year-old. But do they assess each other and base their judgment of whether to fight on the degree of difference between themselves and another buck?

Apparently not, says the Jennings study. They considered the effect of body weight, antler size, age, and prior experience on the duration of fights. They observed that when bucks fought, both participants had some "expectation" that they could win. Bucks with bigger body size or antler lengths were not consistent winners, contrary to at least my expectations. Experience mattered—if a buck had lost to another buck, it was less likely to pursue a long subsequent fight. That explains why we see some bucks just not engage another one. In the end, Jennings and colleagues concluded that bucks of this species based their decision to fight on their assessment of their own abilities rather than how much bigger or smaller they were than the opponent.

How antlers and body size affect interactions between males is one side of the story. We still cannot be sure what visual cues females use when assessing a potential mate, but antler size is relevant. Observing a dominant buck chase off others would give her a clue to her potential mate's status and worthiness as a father of her fawns. A reason for a doe to choose a large dominant buck is that he didn't get that way by luck but rather by his genes. And good genes have a lot of consequences, some not all that predictable.

For example, in 2005 a team of Spanish biologists reported that red stags with large, complex antlers had relatively larger testes and faster sperm than those with smaller, simpler appendages. Hmm, not sure how you race sperm, but I think I'll pass. Not only, then, would a large-antlered male be genetically superior, but insemination would be more likely. Why? After a bull leaves a cow it is possible she will mate again, and the larger buck's having faster sperm reduces the chances that the second buck will father the young.

It is possible then that antler size is dictated by how does choose males. How would you tell? If antlers were only for impressing does, they would not necessarily have to be all that sturdy. John Currey, a biologist from the University of York, compared the strength of antlers and the femur (the leg bone closest to the body), the latter of which ought to be pretty strong. They found that an antler could withstand 2.5 times more sustained force (force attempting to bend the antler) and 6.5 times more blunt force impact than a femur! Fighting, then, seems very likely to be an important function of antlers, solidifying the (at least) dual role of antlers.

8

A NEW KIND OF (UN)NATURAL SELECTION ON DEER ANTLERS
HUNTING

Every year in magazines and newspapers we see a large number of relatively old, mature bucks that were harvested by lucky hunters. Usually my reaction is, "Gee, I wonder where those guys hunt?" The fact that I've never seen one of these brutes suggests that there's some truth to the notion of a once-in-a-lifetime buck. However, after reading a 2009 paper in the prestigious scientific journal *Proceedings of the National Academy of Sciences*, my thoughts took a different turn. Chris T. Darimont and his colleagues review the effects of human predation on deer, especially large males. Typically, we assume that predators remove mostly the young, old, and diseased members of the population. Darimont and colleagues note that hunting exerts a different effect than natural predation and could affect species in new ways. Specifically, hunting often targets (pun intended) the most mature males ("trophies") and removes them from the population before they can breed, or shortens the time they contribute to the gene pool. Does this have a noticeable effect?

Consider elephants. Normally, a large bull elephant has no enemies apart from other similarly endowed bull elephants. But the equation changes when hunting and especially poaching enter the mix. Then, having large tusks, instead of protecting the elephant and making him king of the heap, puts a target on his back. In some areas, a freak mutation, "tuskless," has increased, and many populations now consist mostly of tuskless elephants. In natural history terms, selection now favors bulls with smaller or even no tusks.

Bighorn sheep show a similar trend. In what is probably the best documented case, average horn size decreased about 25 percent over a thirty-year period in Alberta, Canada. No one is positive why, but rams are breeding at an earlier age and with smaller horns. Rams become legal to hunt when they are four or five years old, prior to their peak reproductive age. Removal of the largest rams has apparently resulted in a downward shift in horn size and age at first reproduction.

On the other hand, the revenues generated from hunting bighorn sheep may have aided the population and perhaps compensated for this effect. And no one says that the species will go extinct, because after all, maybe having a population without the largest horn class will not affect the species' long-term survival. However, some researchers studying this issue also think hunting has had other negative effects on the herd. Thus, the problem is this: The genetic quality of a male does not increase with age—he has the same genes at three years of age that he does at six and nine. However, you may not know until a ram makes it to six or more years whether he is truly genetically superior. That's the rub, so to speak.

These trends also are well documented in fish populations. With commercial netting, it is easy to remove the largest fish, and several species now show reproduction at an earlier age and a lack of the largest size classes (which usually constitute the best breeding stock). In other words, the populations are consisting of smaller and smaller individuals. Smaller-sized individuals may produce few offspring or be more vulnerable to predators.

Overall, Darimont and his colleagues concluded that changes in hunted organisms occurred 300 percent more rapidly than changes in nonhunted species. In a summary of a large number of scientific publications, they found that traits such as body and horn size declined in 282 of 297 (94.9 percent) cases, with an average decrease of 18.3 percent. They also found that shifts in reproduction at earlier ages and/or smaller sizes occurred in 173 of 178 (97.2 percent) cases, with an average change of 24.9 percent.

Does this have any bearing on white-tailed deer? If we are removing the most mature whitetails pre-rut, or before they experience their best breeding years, does are being bred by the best of the rest, or the younger, less mature, and possibly less genetically well-endowed bucks that remain. This could reduce the overall genetic quality of the herd.

How, you ask? Well, hypothetically, what if the chances of a buck surviving several bad winters in a row are related to his genetics? If we remove him early because of his large antlers, which signal his good genetics, we might remove the genes that confer resistance to harsh winters. Again, this is hypothetical, but will we eventually see white-tailed bucks without antlers, like the tuskless elephants? A study of deer in Mississippi claims that antler size has decreased, but the harvest pressure was greater than that in most states, including Minnesota. In general determining if overall antler size has decreased in a free-living deer herd is difficult because of the lack of long-term data on the sizes of all bucks taken.

It is unlikely that the trend in the size of bighorn sheep horns will be repeated with white-tailed deer antlers. Whitetails are probably common enough so that enough mature bucks survive each year to ensure the quality of the herd. We do a lot to protect bucks: some states have selective harvest restrictions, antler point restrictions, a one-buck limit, and a hunting season that occurs outside the rut, when bucks are especially vulnerable.

The quality of herd genetics might be enhanced if many of the largest bucks were left to breed. In fact, many of them probably do breed, up to their harvest, but their future contributions to the gene pool are then gone. Now, what if we observed a reduction in antler size in deer similar to that in sheep? Hypothetically, reversing this trend would require a very different strategy than we currently employ.

For example, to protect many fisheries, regulations specify a "slot," or size range, in which any fish caught must be returned immediately to the water. Typically, slots are designed to protect females in their reproductive prime. Imagine a deer season where you could not harvest a buck that scores between 130 and 160 inches. Or you could do it only once in your lifetime. Of course this is hypothetical as you couldn't enforce it or expect hunters to distinguish a 129-inch from a 131-inch buck in the field. But there would need to be a way to get some bucks through the slot to provide enough mature bucks to maintain superior genetics.

In the end, I think our efforts to enhance the habitat and survival of deer probably outpace whatever negative effects occur to antler size because of removing the prime reproductive males. At least the photos of huge-antlered bucks in magazines suggest that. But I'm glad deer are not like bighorn sheep, and I'll think twice about harvesting even a nice two-year-old buck. The does are better eating anyway, especially the younger ones, which raises the question of which does we should harvest, a topic for another time.

9

MY DEER DOCTOR
TAKE TWO ACORNS AND CALL ME IN THE MORNING

To my knowledge, I have never seen a deer that was "under the weather," in the same way that you can tell someone has a cold or flu. I've not seen a deer just lying around looking crummy, sneezing and whiffling, sitting next to a pile of used tissues. This observation leads me to wonder if deer get sick. That curiosity led me to do some digging about what illnesses and parasites deer get. It turns out that if you wanted to be a deer doctor, you would be expected to know about viral, bacterial, and rickettsial diseases as well as parasites like protozoa, trematodes, nematodes, cestodes, and arthropods. Pretty much you would need to be up on the diversity of life and be an expert in treatment of disease. Much of what follows comes from T. A. Campbell and K. C. VerCauteren's chapter "Diseases and Parasites" in the book *Biology and Management of White-Tailed Deer* (ed. D. G. Hewitt, CRC Press, 2011).

Deer get Chronic Wasting Disease (CWD) and bovine tuberculosis. These two diseases are very different. CWD is caused by a naturally occurring protein that changes its shape (then called a prion) and interacts with other normal proteins, causing them to change their shape; they build to a level sufficient to clog the central nervous system, especially the brain, resulting in death. Prions are not like other disease agents—they are not alive, they don't mate or reproduce—they are more like an environmental toxin. That means that making a vaccine to protect against prions may not be possible, any more than making a vaccine to protect against cyanide poisoning.

Tuberculosis is a more typical disease in that it's caused by a bacterium. It occurs in humans and is one of the most common infectious diseases worldwide. Bacteria can be killed by antibiotics, and we've done a good job of controlling tuberculosis in people. Because a bacterium has genetic material and is capable of reproducing, it is very different from a prion. It can evolve so that it is resistant to vaccines and antibiotics. You get a new flu vaccination every year for the same reason—the influenza virus evolves so rapidly that last year's flu shot may not provide immunity. The main concern with bovine TB is that it can be spread by oral or respiratory routes, such as via contaminated feed, mutual grooming, or inhaling infected droplets from sick animals. This disease is very bad for both deer and cattle.

A nasty viral disease is hemorrhagic disease (HD), which is caused by a special virus called an "orbivirus." There are more than 120 "strains," which fall into fourteen major groups, but only a couple are really bad for deer. The disease is spread by several species of midges (so you have to be an entomologist). HD is the most important viral disease of deer, but population-level effects are not well documented. Also, it's not zoonotic, meaning it is not transmissible to humans, so it hasn't received our undivided attention. Deer with clinical expression of HD can exhibit depression (you have to be a deer psychiatrist too), emaciation, facial swelling, fever (thank goodness for the old thermometer), lameness, loss of appetite, reduced activity, and respiratory complications. So if you have a mildly sick deer, HD is a possible cause, and you have to do a battery of blood tests to confirm the diagnosis; if you don't want to do all that testing, you might use the "take two acorns and call me in the morning" diagnosis. Deer with really bad (gross) lesions could have HD, likely in the later stages, but to be positive you have to find the virus in the deer. Better hope they have good insurance.

Six types of virus causes hairless tumors, or warts (technically speaking, cutaneous fibromas), which can be anywhere on the body and are usually smooth—although sometimes they look like a head of cauliflower. West Nile can affect deer, but there are few known fatalities. Another virus causes symptoms that look like foot-and-mouth disease, including fluid-filled blisters on the mouth, tongue, muzzle, teats, and feet. Sounds like my old VW bus—what's wrong could be lots of things.

Moving on to the bacterial world, we are not sure if deer can get anthrax. Some herbivores get it, but it doesn't seem to affect deer—at least no deer with clinical signs of anthrax infection have been reported. A skin disease called dermatophilosis sometimes infects whitetails, especially fawns, and is recognized by hair loss, thick scabs, and emaciation. Typical lesions are raised, matted hair tufts, held in place by a crust of shed skin and pus. Right there is the reason I didn't become a medical doctor—pus.

A bacterium called *Arcanobacterium pyogenes* causes lack of coordination and fear, blindness, weakness, profound depression (as opposed to just depression?), emaciation, circling, lameness, fever, and anorexia (maybe you can work on skinny movie stars too?). Those familiar with CWD will recognize all these symptoms. The reason is that the bacterium causes a brain abscess, the effects of which mimic the accumulation of prions in the brain, causing the same symptoms deer with late-stage CWD have. This is why it is not possible to know that a deer with these symptoms has CWD; only microscopic examination of the brain can distinguish these two very different diseases.

Other bacterial diseases of whitetails include Johne's disease (paratuberculosis), salmonella (occurs mostly in fawns), and Lyme disease. Like dogs, deer also can get anaplasmosis, a rickettsial disease, transmitted mostly by ticks. The biggest issue is that deer, especially mule deer and blacktails, can carry the bacterium and spread it to cattle; the mortality rate for infected cattle reaches 50 percent.

So, you've mastered these and a host of other diseases, but what about parasites? One of my favorite biological "truisms" is this: the most common substrate for an animal to live on is another animal. It makes sense when you think about it. Most of the animals we know, like deer, have parasites. Parasites themselves often have parasites, and these parasites can have parasites. If you add it up, the truism is true! Now, my deer doctor, your education needs to expand even further.

Deer provide homes to a surprising number of parasites, including protozoans, trematodes, nematodes, cestodes, and arthropods. Deer observers rarely see these creatures, but they are important to deer well-being nonetheless. I'll mention but a few.

Deer harbor several kinds of parasitic protozoans (single-celled organisms), and they cause diseases like toxoplasmosis, babesiosis, and theileriosis. These generally do not do great harm to deer but are cause for some concerns. For example, in some areas 30 to 60 percent of deer are carriers of toxoplasmosis, and it can be transferred to humans (with bad consequences) if the venison has not been frozen or is undercooked (and never frozen). Neither babesiosis or theileriosis infect humans, but a form of bovine babesiosis has been found in deer in northern Mexico, which concerns southern ranchers.

Liver flukes infect deer but are not a problem for human health. A whitish "large lungworm," which can be up to 1.5 inches in length, builds up in the lungs and causes serious problems to deer, but they seem to infect mostly fawns. Larvae can be found in deer feces, and other deer pick up the larvae that crawl into vegetation, so the life cycle involves only deer. In areas where there are too many deer for the available food supply, lungworms are a serious cause of deer mortality. Humans are not at risk from deer infected with lungworms.

Many deer host the "large stomach worm" without showing any outward or clinical symptoms. In some areas, like the southeastern plain of the United States, up to 100 percent of the deer have this parasite. A very high infection rate is indicative of a herd that is exceeding its carrying capacity. In other words, animals that are stressed from lack of nutrition and overcrowding are susceptible to large stomach worms—one of nature's ways of compensating for overpopulation. These worms are not a problem for humans.

The meningeal worm is common, and the white-tailed deer is the definitive host (meaning it is where the worm reproduces). Like those from the bacterium mentioned above, symptoms can be confused with those caused by CWD. This worm has a complex life cycle that involves terrestrial mollusks as intermediate hosts, which deer inadvertently consume, thereby infecting (or reinfecting) themselves. Although this worm is not a threat to human health, all other native cervids (e.g., mule deer) are susceptible to this worm.

The adult arterial nematode worm lives mostly in the carotid arteries of mule, white-tailed, and black-tailed deer. A deer with this worm shows symptoms called "lumpy jaw." Several species of tapeworms call deer home. Only rarely do they cause the deer to show clinical signs of illness. The tapeworms get passed back and forth from herbivores to carnivores.

To this point, relatively few people will have recognized any of these parasites. But many have found ticks on deer. Ticks are called ectoparasites, meaning they live on the outside of the body. Ticks rarely cause severe problems for deer. Probably the most familiar is the deer tick, which is a problem for humans because high deer numbers mean more ticks, and they vector several diseases, including Lyme disease.

Deer also get ear mites, mange mites, nasal bots, louse flies, sucking lice, and chewing lice. Nasal bots are particularly disgusting creatures (I guess not to themselves), but none of these pose major threats to deer.

Lest the reader become bored with this academic laundry list of parasites, let's think about the broader consequences and implications. The old professors (of which I'm now one) who told me that the most frequent home of an animal is another animal knew their stuff. But this leaves the question, why don't we see a lot of these creatures? First, many are endoparasites, not easily observed without close inspection of the insides of a deer. Second, most parasites live by a "code of prudence." If a parasite infects its host to the degree that it dies, and its reproductive success (evolutionary fitness) depends on its young being dispersed, the parasite has failed. A parasite benefits from its host being infected but still able to move around spreading the parasite's offspring. In those cases where severe infection does debilitate an individual deer, it is killed quickly by predators or its carcass is eaten rapidly by scavengers. So, you are not likely to see a deer that is so sickened as to be debilitated, because it doesn't last long in nature.

Lest you feel sorry for deer, don't forget that humans have easily as many parasites as deer. And humans have some doozies, like the worm in Africa that eats the eyes from the inside out, and the guinea worm (probably in fact the biblical "fiery serpent"), which you should read about but not right before a meal. Maybe deer don't have it so bad after all.

10

TRYING TO OUTFOX DEER TICKS AND LYME DISEASE

When I taught the University of Minnesota field ornithology class at Lake Itasca, I was always fond of the times when wood ticks were out in force. I would tell the students, as they were squirming to find a tick (or one they imagined), that I love ticks because there are plenty of them and they're free. What else can you say that about? I personally don't mind all that much if I have ticks—I kind of like finding them. But that isn't true for everyone. Ticks also fascinate me because they can bring such a change in behavior in some people. Even the most die-hard vegan, antihunting student can turn killer when confronted with a tick. One took to stabbing them with a pen on his desk, watching the tick's legs flailing about as it was pinned to the desktop, while maintaining a strict vegan lifestyle (the student, that is). I guess we all have our limits, and finding a tick crawling on our body provides a short track for many to discovering personal boundaries.

I do admit to checking if the ticks I find on me are wood or deer ticks. As we all know, deer ticks vector Lyme disease, which is a bacterial infection that can lead to serious health issues in humans (and dogs) if left untreated by antibiotics. The disease was not recognized by health care professionals until about two decades ago, when the tick–disease link was discovered. We still hear of cases today where the disease has gone undiagnosed until the later stages, partly because the red "bull's eye" doesn't occur in all infected people, deer ticks can be hard to find and so people don't realize they were bitten, and the symptoms are mimicked by a number of other illnesses. But what role do deer play in the transmission of Lyme disease, given that a principal vector is called the *deer* tick? A 2012 scientific paper by Taal Levi and colleagues in the *Proceedings of the National Academy of Sciences*, one of the leading scientific journals, addressed this issue from a theoretical and modeling perspective, with a good dose of natural history.

The authors reported that Lyme disease has increased in many states, including a whopping 380 percent increase in Minnesota, and 280 percent in Wisconsin from 1997 to 2007. But the winner is Virginia, with an increase of 1,300 percent. Many people have assumed that the increase in Lyme disease is a result of the increase in the white-tailed deer population, beginning a century ago. As I noted in a different essay, the deer population in North America is currently at or near an all-time high. But exactly how do deer figure into Lyme disease transmission?

Deer do not transmit the bacterium that causes Lyme disease to deer ticks, but they are important in providing blood meals for ticks of all life stages, allowing them to reproduce in large numbers. Most (80 to 90 percent) ticks get the disease from one of four fairly common small mammals: the white-footed mouse, the eastern chipmunk, the short-tailed shrew, or the masked shrew. Given that deer abundance peaked and stabilized long before the recent increases in Lyme disease, as the authors noted, the idea that the increasing prevalence of Lyme disease is due to having lots of deer might not make sense. They considered an alternative explanation.

The authors think that an interaction between larger predators and the small mammal vectors of Lyme disease mentioned above might explain the recent increases in cases of Lyme disease.

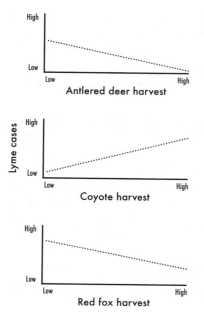

Interconnection between numbers of deer, coyotes, and red foxes and the incidence of Lyme disease. *Top:* As the number of antlered deer harvested increased, the incidence of Lyme disease decreased. *Middle:* When more coyotes were harvested, Lyme cases were more frequent. *Bottom:* When more red foxes were harvested, Lyme cases decreased. When red foxes are numerous, they keep rodents that spread Lyme disease in check. Coyotes eat red foxes, so when there are more coyotes, there are fewer red foxes and hence more rodents, and the number of Lyme cases goes up. Deer serve as hosts for the Lyme-carrying ticks, so more deer (lower harvest) means more Lyme cases.

They plotted the relationship between the number of Lyme cases and harvest numbers of antlered deer, coyotes, and foxes. In the accompanying figure I show the patterns for Minnesota.

The plots show the following. There were fewer Lyme disease cases when the harvest of antlered deer was relatively high (top panel). This finding does not fit with the traditional idea that high deer densities are responsible for high Lyme prevalence. In fact, studies show that where deer densities were reduced on purpose, the percentage of infected young ticks did not change. The middle panel shows that a lot more Lyme cases occurred when more coyotes were harvested, presumably meaning there were a lot of coyotes around (and not just more coyote hunters and trappers). The lower panel, perhaps the most telling, shows that when there were fewer red foxes harvested, likely meaning there were fewer red foxes, a lot more cases of Lyme disease occurred. These trends hold true in three other states (Wisconsin, Pennsylvania, and Virginia).

How do these observations help understand recent increases in Lyme disease? Well a little bit of natural history detective work suggests a fairly simple explanation. More coyotes means there are fewer red foxes, owing to coyotes killing foxes. Foxes, however, are natural predators of the chipmunks, mice, and shrews, the animals that are the primary vectors of the Lyme bacterium to the deer ticks. So, if there are fewer foxes, then there are more small mammal hosts for the bacterium, and you get a big bump in the number of infected larval ticks that fall off the small mammals and then attach to you or your dog and transmit Lyme disease. (Deer ticks take three blood meals from successive hosts, but we think that only the first stages can obtain the bacterium by feeding on an infected small mammal.)

It is still true, however, that large numbers of deer aid the tick population by providing feeding grounds for deer ticks. That is, lots of deer translate into lots of meals for ticks, even those not carrying the Lyme bacterium, but their offspring might latch onto one of the small mammal vectors.

The authors suggest that this is an interesting case in which human alteration of the predator community has cascading effects that influence the transmission of Lyme disease, via an unanticipated pathway. The authors propose that experiments manipulating densities of deer, coyotes, and foxes might be a good way to test their models and conclusions. Clearly they have come up with an intriguing idea, and the goal now is to see if it holds up to field study. One would think that surveys of the small mammals should also be done to see if there is an inverse relationship between their populations and fox densities.

I do not preach to others about how or what to hunt. That said, at this point in time I myself would not consider killing a red fox. They are providers of an important "ecosystem service," by controlling small mammals that harbor the Lyme bacterium. On the other hand, foxes are predators on nests of ground-nesting ducks, and more foxes might mean fewer ducks. Of course, foxes and ducks coexisted at an earlier time when there were far more ducks, so a preferable goal might be to better balance the natural densities of coyotes and foxes. With fewer ducks today, maybe foxes are too efficient a predator.

A course of action to reduce Lyme disease incidence, then, is complicated by lack of knowledge of the normal states of predator and prey densities and potentially competing management interests. If Levi and his colleagues are right, more foxes could lead to reduced levels of Lyme disease. But more foxes might mean fewer ducks. Which should we prefer? The recent increases in Lyme disease rates might well be a cry from nature that the predator community is out of whack and needs to be brought back into line. Until we can figure out what "back into line" means, be sure to watch for Lyme-disease symptoms in yourselves and your pets and support researchers trying to figure out how to raise more wild ducks.

11

DEER AND THEIR SUBSPECIES
FACT OR FICTION?

Anyone who has looked at more than one white-tailed deer in an area has usually noticed at least subtle differences between them: darker or paler, different patterns of white, different sizes even at the same age and sex. We are very attuned to seeing differences among people, even from the same family. If we were as astute in seeing (and smelling) deer, we would likely think they are just about as different as are kids in a family, and as are you from your aunts, uncles, and grandparents. Such is the nature of heredity—variation is the nature of life, and the closer you are to someone in your genealogy the more similar you tend to look.

These sorts of comparisons can be taken a step further by comparing how deer look in one area with those that live elsewhere. Just about any organism that is distributed over a broad area shows some sort of regional differences, a sort of bar code linking them to their specific home area. The phenomenon is called geographic variation. Typically we think that geographic differences are related to how plants and animals adapt to geographically varying environments. Over longer periods of time, we think that geographic differences lead to new species.

Another activity that students of geographic variation have engaged in is providing taxonomic names for recognizably different populations in different areas. Most readers know that animals have a genus and a species name in Latin, as part of a system of classification devised by famous eighteenth-century Swedish scientist Carl Linnaeus. Thus, when someone tells you they observed an *Odocoileus virginianus*, you know that they saw a white-tailed deer. The classification system is designed such that you know that other deer species in the same genus share more features in common than they do with species in other genera.

Taxonomists (the namers) have also extended the Linnaean

system below the species level to a category known as subspecies, which correspond to morphologically distinct populations living in different areas. Incidentally, for the record, *species* and *subspecies* are both singular and plural; thus, *subspecie* is not a word. By classifying the population as a subspecies, taxonomists call attention to the observation that these populations differ in some way from geographically more distant populations. For example, within species of warm-blooded vertebrates, individuals living in the north tend to be largest. We think this is because when body size increases, surface area increases as a square, but volume increases as a cube; therefore, as an animal increases in body size, it will lose less heat through its relatively smaller surface area and be better adapted to cold places. White-tailed deer are a good example (think of the pictures of huge-bodied Saskatchewan deer). This general phenomenon is called Bergmann's rule, although there are exceptions.

Back to subspecies. Although the species category is relatively robust, the same cannot be said for the subspecies category. The reason is the lack of consistent criteria for recognizing subspecies. Some confounding factors are the following: How many geographic areas were studied? How many specimens were examined? How many features were studied? How much do they vary? Is the variation geographically discrete or continuous? Thus, a subspecies of one animal might be an arbitrary grouping based on subtle differences in coat color, whereas another subspecies might be a very clear and geographically distinct division. Some taxonomists adhere to a "75 percent" rule, where 75 percent or more of the individuals must be recognizable as one or another subspecies. However, most subspecies do not match even this goal. The standards are often lax, and it sometimes seems that there were no rules applied at all. To note an extreme example, the subspecies O. *v. leucurus*, from the Columbia River area, was based on a single individual that was not preserved but eaten; typically all scientific names are traced to a museum specimen so that it can be studied by others and serve as a perpetual reference.

The subspecies category can be a total mess depending on

the species. The white-tailed deer is, in my opinion, such a species. This sentiment was echoed in a book chapter by James Heffelfinger (in *Biology and Management of White-Tailed Deer*, ed. D. G. Hewitt, CRC Press, 2011). He notes that across its extensive range, thirty-eight subspecies of white-tailed deer are currently recognized (by someone), although at least sixty-five others were described in the past. This large number reflects the attention of many independently working taxonomists using different criteria and the fact that whitetails occur not only in North America but south into South America. Further complicating the situation is the large amount of variation between the extremes, and that figuring out where to draw taxonomic lines for subspecies is basically arbitrary.

Minnesota is a good example. There are reputed to be three subspecies (see accompanying figure): one corresponding in location in part to the prairie (*Odocoileus virginianus dacotensis*), another corresponding to the boreal forest and big woods (*O. v. borealis*), and a third subspecies (*O. v. macrourus*) that extends from Minnesota in a narrow strip south to Louisiana.

Does this mean that if you harvest a deer from each of these three subspecies, you need to expand your trophy room? Actually, it doesn't. These subspecies of white-tailed deer are examples of applying discrete names to a gradual pattern of variation—they are not geographically or morphologically discrete.

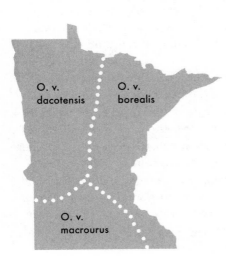

Approximate boundaries of subspecies of white-tailed deer in Minnesota.

So if you harvest two deer, each five miles on either side of the subspecies division, they will be more similar on average than two from the same subspecies taken five hundred miles apart. Although we have not done a genetic study, I bet that I could not tell from the DNA what subspecies you harvested.

What have others said? Heffelfinger said, "Minnesotans do not spend much time thinking about the three different whitetail subspecies designated in their state, nor should they." At a larger scale, if discrete differences among the subspecies really existed, you would think they would be recognized by the Pope and Young Club and the Boone and Crocket Club. However, only the Coues white-tailed deer, from desert regions in Arizona, New Mexico, and western Mexico, is recognized.

Are all deer subspecies so poorly demarcated? Almost. Besides the Coues deer, the Key deer, isolated on islands off the southern tip of Florida, is morphologically and genetically different. This might be a result of the fact that the northernmost populations are separated from the mainland by twenty-five miles of open ocean. Key deer, which number probably less than one thousand individuals, are on the U.S. endangered species list, and no official categories exist for (past) trophies.

To summarize subspecies of white-tailed deer, imagine a contour map of North America, with rather low peaks and shallow valleys representing size, coloration, or some other feature that transition gradually from one peak (subspecies) to another. At one extreme are the diminutive Key Deer, and at the other, the very large-bodied individuals from the far north. From this contour map, we can easily see the difficulty of applying discrete names to represent subspecies. In fact, there would be nearly infinite ways to assign names to the variation, or at least sixty-five. There are probably only three "actual" subspecies of white-tailed deer: Coues, Key, and everything else. Further study might reveal greater numbers in South America.

We are stuck with this historical legacy. The U.S. Endangered Species Act affords protection to "endangered or threatened" subspecies, irrespective of how distinct they are, simply because they have a taxonomic name. That's unfortunate given how few of the thirty-eight are truly distinct. However, there is some hope. Although few biologists today actually name subspecies, many subspecies are tested with modern genetic methods, and we are finding that only a small portion of subspecies are genetically discrete. We hope that taxonomic names will someday reflect more meaningful divisions of species in nature.

12

CAN GAME MANAGERS CONTROL THE NUMBER OF DEER?

In some areas, the deer herd is out of control. This is especially true in some eastern states and in many rural or metropolitan areas, where abundant food and lack of predators have raised numbers to well beyond what the habitat can support. Habitat destruction, damage to crops and ornamental plants, and deer-vehicle collisions are all a result of overabundant deer.

For many game species, managers often adjust seasons and bag limits to reduce the harvest of certain age and sex classes. Some states specify slots for fish, roosters-only take for pheasants, and limits on the number of hen mallards that can be harvested by duck hunters. These are all designed to keep populations at a level that will sustain harvest, by protecting vulnerable sex and age classes deemed most important for reproduction.

In the case of deer, various types of strategies are used to manage the herd. Some states use antler-point restrictions (APR), designed to yield greater representation of older bucks. There is merit to this, as often older bucks have "proved" their genetics and are the ones we want siring fawns; if they're killed as one-and-a-half-year olds, they will not get the chance. We have also seen the use of special seasons or intensive-harvest areas for does. The most effective way to control the growth of the deer herd in areas where there are more than can naturally be sustained is to reduce the doe population. Alternatively, when there are fewer deer than desired, only bucks are allowed to be harvested, owing to the fact that relatively few bucks father most of the young. Managers must balance the harvest, area by area, to reach an appropriate level of deer abundance (which depends on which stakeholder group you ask).

The Wisconsin Department of Natural Resources decided to try to manage deer via its Earn-A-Buck (EAB) program. Basically, a hunter had to register an antlerless deer to be able to fill a buck tag. I heard quite a few complaints, and one of my Wisconsin hunting friends even asked if I knew where he could "borrow" a Minnesota doe. There was relatively little support in Wisconsin for the EAB program and the October antlerless hunt (which get bucks spooky before the gun season), and they were discontinued in 2011. In some areas the population size was at the preset goal and didn't need to be reduced by continued pressure on does. In areas where the population is still considered to be too high, lots of antlerless tags were made available.

A scientific assessment of the EAB program was published in 2010 in the *Journal of Wildlife Management* by Timothy Van Deelen from the University of Wisconsin-Madison and colleagues from the Wisconsin DNR. In addition to the EAB program, they also considered the use of opportunity-based supplemental antlerless-only seasons (SAOSs). For example, there were four-day (December only) and eight-day (four days in October plus four in December) SAOSs. The effects were judged at the level of Wisconsin's Deer Management Units (DMUs), which are areas delimited years ago and used to monitor populations and harvest. In addition, they factored in variation in the amount of forest or farmland habitat, deer and hunter densities, winter severity the previous year, and amount of public land available for hunting. As one can appreciate, figuring out the effects of programs like EAB is actually rather complicated unless you know these things. For example, fewer bucks might be harvested in a year during which EAB is in use, but maybe there were just fewer deer in the area that year, the previous winter was unusually harsh, or there were fewer hunters. One also needs information from many years and a large geographic area to assess the program.

The data were from 1996 to 2008 and included 107 DMUs. The study included 41 four-day and 205 eight-day SAOSs and around a hundred seasons when EAB was required. Mean pre-hunt deer density was 6.9 deer per square mile, and mean post-hunt density was 4.6 deer per square mile. Density of hunters on opening day of the firearm season averaged 2.8 per square mile.

The results were not completely surprising. Harvests of all deer increased with increasing deer and hunter densities, as you would predict, but were unrelated to winter severity the prior winter (which I wouldn't have predicted). Fewer deer were harvested in areas with more public land open to hunting—I thought the opposite would be true, but maybe all the pressure pushed them to private land. In fact, the researchers noted that "at high enough densities hunters interfere with each other and deer become more vigilant as hunting pressure increases, suggesting a point at which additional hunters may become counter-productive to management goals designed to increase harvest."

The SAOSs and EAB programs worked. Take of antlerless deer increased owing to the extra antlerless seasons by almost 50 percent. EAB resulted in harvest of an extra 0.75 antlerless deer per square mile. In fact, the extra seasons and EAB worked together. The effectiveness of the SAOSs more than doubled when coupled with EAB, with something up to a 139 percent increase in harvest of antlerless deer in some areas. These results were important, as one study suggested that as many as 50 percent of hunters would not have taken an antlerless deer without the need to comply with EAB.

As many would have predicted, the harvest of antlered bucks also decreased. This is good or bad depending on your point of view. If hunters had limited time afield and were required to take an antlerless deer, they might not have had enough time to hunt a buck later. More hunters harvesting does might have contributed to bucks becoming even more nocturnal and more wary. This contributed to hunter frustration with the EAB program. The researchers also thought that perhaps since hunters had already put some venison in the freezer, they might have been more selective and waited, unsuccessfully, for a really large buck. The researchers noted that buck harvest didn't change when only the two short antlerless-only seasons were used to reduce the herd size.

A friend reminded me that a potentially unintended consequence of EAB and the SAOSs was the harvesting of buck fawns, called button bucks. If a hunter had a limited time to hunt, he or she was unlikely to have discriminated between doe and buck fawns (which count as a "doe"). Given that perhaps half of all hunters wouldn't have harvested an antlerless deer, it stands to reason that quite a few buck fawns were harvested as part of the EAB program that otherwise would have survived their first fall.

I was struck by the fact that I had not thought seriously about how to manage deer populations. I read many letters to editors that make it seem simple: either "do x" or "don't do y" and you'll have more big bucks. But even something seemingly as simple as judging the effects of EAB is far from simple. There are many potentially confounding pieces to the puzzle, and we need good long-term data, and then the task is to consider the effects simultaneously. This means that what appears to be a clear, simple result is often clouded or disappears altogether.

Thus, it would not be fair to state that EAB didn't work. The two programs, especially when used together, have the potential to change the number of deer and to shift the sex ratio to fewer females and relatively more males. If the goal is to reduce the size of the herd, these two programs are effective. However, a management authority (e.g., a department of natural resources) is also interested in maintaining hunter satisfaction, and the researchers noted that the results of EAB and SAOSs would be viewed as a positive outcome by some but not by others. The Wisconsin DNR seems to have done a good job of balancing both. Perhaps the message to deer hunters is that if you don't want an EAB in your area, harvest does when the DNR requests you to do so.

13

MOUNTAIN LIONS, PRIONS, AND SICK DEER

Chronic Wasting Disease (CWD) is a neurological disease that produces small lesions in brains of deer (white-tailed, mule), moose, and elk. Infected animals are in poor body condition, exhibit behavioral abnormalities, and later die. Infected deer apparently do not recover from CWD, but studies suggest that some deer may have a genetically based partial resistance (see earlier chapter on CWD). The disease is similar to scrapie in sheep and mad cow disease in cattle. Most diseases we are familiar with are caused by bacteria or viruses, but CWD is likely caused by a naturally occurring protein that for a mysterious reason becomes misfolded and is then called a prion, which is infectious and can destroy its host. Prions build up in nerve tissue, causing the death of nerve cells and loss of normal body functions. Confirmation of CWD is obtained by examining the brain tissue from a dead animal under a microscope and finding spaces (holes).

Attention to CWD waxes and wanes in the popular press depending on the occurrence of outbreaks or new findings of spread. Nonetheless there is still considerable scientific interest in this disease and in understanding its ecology and effects on deer (and elk and moose) populations. The cause for the attention given to CWD is at least twofold. First, it can severely affect a deer or elk population, and deer hunting is a hallowed pastime as well as a major economic factor. Second is the possibility that the disease can be transferred to humans who consume infected deer. Because CWD is like mad cow disease, it garners our attention.

However, the Centers for Disease Control found no evidence that the disease has ever crossed the "species barrier" and infected humans. Still, we are advised against eating infected deer, cutting through bone, and eating certain organs, like brain and spinal cord, that seem to harbor the highest concentrations of prions.

So, that brings me to the actual point. Animals in the wild get sick. And we all know that in an ideal world, predators, such as mountain lions, preferentially kill the sick and the old, thereby keeping the herd healthy (the so-called sanitation effect). Actually, I doubt that a mountain lion would pass up an easy kill on a healthy deer. But not a lot of evidence exists that confirms that such predators preferentially take out the old and the sick, like those with CWD—it just sounds reasonable. A recent paper on mountain lion predation on mule deer suggested that maybe predators behave like we think they should. How the information was obtained is pretty cool.

As anyone who has spent time outdoors knows, locating a lot of mountain lions, observing them kill a lot of deer, and then figuring out if the lions were preferentially killing sick and old deer are not easy tasks. But in the mountains of the northern Front Range in Colorado, Caroline Krumm and her colleagues did the next best thing (*Biology Letters* 6: 209–11 [2010]). They captured nine mountain lions and fitted them with GPS collars. Then, they looked at the GPS coordinates transmitted from the collars and predicted that a cluster of locations was where the lion had made a kill. They went out and found the kill, and if it was a mule deer, they took samples to determine if it had CWD. Talk about armchair biology!

They found 54 lion-killed mule deer carcasses two years or older (since younger deer might not show signs of CWD infection, younger lion-killed deer were not used). Of these, 12 had CWD (22 percent), and there was some indication that bucks were more likely to be infected than does.

So, you might think that lions seek out CWD-infected mule deer. However, to be able to say this, you would have to know what the prevalence of the disease is in the general population. Well, hunters were able to come to the rescue. To answer the question about baseline infection, Krumm and colleagues sampled 312 hunter-killed deer and found 23 (7 percent) with signs of CWD.

Specifically, they concluded that "hunter-killed female deer were less likely to be infected than males, but both female and male deer killed by mountain lion were more likely to be infected than same-sex deer killed in the vicinity by a hunter." Now, of course, maybe hunters are more likely to kill CWD deer because they are less wary, and in the early stages of the disease, hunters don't notice the symptoms. After all, a deer that presents an easy shot is hard to pass up, and the clinical signs usually appear late in the infection, near the deer's death.

So, it's not absolutely clear what the actual level of CWD is in the general population, but it's higher in lion kills than hunter kills. In a related study, samples were taken from road-killed deer, and the researchers found that prion-infected deer had a greater chance of being killed by a mountain lion than by a car. Of course, a deer with CWD might be more likely than a healthy deer to stray in front of a vehicle. Given the number of road-killed whitetails, deer are not particularly savvy when it comes to crossing the street. Figuring out what the real baseline level of infection is in wild deer isn't easy, but the roadkill data come pretty close.

Krumm and colleagues summed up by concluding that "mountain lions may also learn to recognize and more actively target diseased deer." As the lions don't get sick from eating CWD-infected deer and they're probably pretty easy to catch and subdue, it makes sense. And there is an added potential benefit: the lions eat most of the carcass, including the brain, which may prevent further spreading of prions to uninfected deer. So, maybe we should be managing our lion populations, so that if CWD becomes widespread (which we hope is never), we will have a natural sanitizer available.

14

THE RUT
(MAYBE MORE THAN YOU WANTED TO KNOW)

Deer hunters all nod their heads knowingly whenever someone mentions "the rut." The word instantly evokes vivid memories of a cold time and the almost unbelievable sight of a mature white-tail buck walking around in broad daylight, something you almost never see the rest of the year. Most people know that these bucks are searching for does in estrus so that they can exercise their primal urge to procreate.

But like so many familiar things, a bit more thought reveals some less well-known aspects of the rut. Why is it so concentrated in time? Why aren't female deer receptive all year long like humans? Do does ever mate with more than one male, and if so, does it "matter"? If you're a twin, does your sibling have the same last name as you? How many young in a given area are fathered by the dominant buck? Does every buck have, so to speak, a fighting chance at being a dad?

The question of why does aren't receptive year-round has a couple answers. First, if does were receptive year-round, bucks would have to be antlered all year. There would be no point in dropping and regrowing antlers, and the entire buck biology would change. Second, if you're a fawn in Minnesota, being born in the spring gives you the best chance of survival over any other time of the year.

But why is the rut so concentrated in time? The reason is called predator swamping. A fawn is a pretty helpless creature and a great meal for many predators. However, no predators prey exclusively on fawns. Why? Because does are in heat for a short period, it results in lots of fawns being available only for a very short time (swamping). If a predator evolved to specialize in finding and eating fawns to the exclusion of other prey, it would be hungry most of the year and starve. As a result, there are no "fawn-specialist" predators, although fawns are taken by many "generalist" or opportunistic predators.

What about the does we see running from amorous males (reminds me of bar scenes from my long-ago youth)? Obviously they eventually choose a mate. But once they've made that choice, do they mate with other bucks? And if they do, what are the consequences?

One could watch does and see what they do, but they're too mobile to make this a worthwhile strategy. The obvious choice is to use DNA technology, where paternity and maternity leave their marks in a bar-code sort of way. Randy DeYoung, at Texas A&M University-Kingsville, did just that.

His group's work challenges the notion that dominant bucks gain most of the matings in a given area. They found that in captive and wild herds, the largest buck is usually dominant, as we suspected, but sometimes he's not, and in any event the dominant buck does not father all the local offspring. The data show that the most successful bucks mate with six to seven females in a single year. But many bucks less than three and a half years old father some fawns, in fact, 30 to 33 percent of all fawns in an area. Why? I thought the dominant bucks ran these guys off.

The explanation is that when a big buck is busy tending an estrous doe, for up to twenty-four hours, the subordinate bucks mate with other estrous does. A dominant buck will stay with an estrous doe and defend her from other bucks (at least we got that right), rather than risk leaving her and having another buck breed her. Why he might do this is now pretty clear from DeYoung's genetic work. The genetic tests showed that in cases of twins and triplets, up to 25 percent of the litters had two dads. Obviously, then, if two bucks mate with a doe, each might be successful. This might not happen if a dominant buck tended the doe until she was no longer receptive. So it "pays" for a dominant buck to tend an estrous doe rather than mate and move on. I guess you could call it a doe-in-the-hand strategy.

What is not known is whether females that have a single fawn mated with multiple males. And in cases of twins with the same father, did she actually mate with more than one male (and one was just unsuccessful)? In both cases, what determines which of the males will father the fawn(s) is not clear. The fact that bucks apparently don't leave estrus does might mean that the last male to mate actually fathers the most young. Maybe if she mates a second time after twenty-four hours, it doesn't "work." Now of course we should not overlook or discount the doe's strategy (why would she seek additional matings?), but we'll leave that to later.

The name of the game, evolutionarily speaking, is fitness. You win if you leave more offspring. If a buck mates with multiple does, they can leave more offspring than a single doe. That is, a doe might have twins or rarely triplets, but if a buck mates with six or seven does and at least some have twins or triplets, his genes get into way more fawns than if he focuses on one doe. So, bucks try and breed as many does as possible, whereas does in theory are "choosy" and pick the genetically best male they can find (how they do this is another story).

And now for the truly esoteric. What if, with emphasis on the *if*, bucks could produce more Y-bearing sperm than X-bearing sperm. Then the does he breeds have a greater chance of producing buck fawns than doe fawns. It might be in a buck's best interest if his does produced more buck fawns than doe fawns, as buck fawn sons will eventually have greater "fitness" (more offspring) than his doe fawns because of the mating system.

Sounds wild. How could this occur? We know that males of some animal species can produce sperm that are Y-chromosome biased; that is, their sperm will yield more male offspring than female offspring. How would you find out whether this occurs in white-tailed deer? DeYoung and his colleagues collected semen from captive (anesthetized of course) white-tailed bucks. Using "fluorescence flow cytometry," they found that bucks produce a fifty-fifty mix of X- and Y-bearing sperm, meaning that bucks are not "slanting the odds" toward producing buck over doe fawns.

Despite the amazing things that whitetails do, they appear to be just a regular mammal in this regard.

Some fear that trophy hunting is changing the genetics of deer herds and reducing antler size by removing the genetically best and dominant males. This is based on the notion that the dominant buck fathers most of the fawns in a given area. It now appears that the dominant buck does not do the majority of the mating in a given area. That might explain why deer antler size is apparently not decreasing over time to the same degree that bighorn sheep horns are decreasing. Researchers like DeYoung, using modern molecular tools, are giving us cool glimpses into the mating systems of many animals, including whitetails. We have watched white-tailed deer for decades without actually understanding what was going on. I am waiting for the next set of studies that will hopefully answer questions like what determines who actually fathers the young in multiple matings. And he promised to tell me whether a given doe and buck breed together for more than one year. Or is it love 'em and leave 'em?

IN THE WOODS

15

HUNTING SPOTS FOR WILD TURKEYS
AT THE LAST GLACIAL MAXIMUM

Glaciers have been a part of the long-term history of Minnesota and the entire northern part of North America for the past 2 million years. During cold periods, glaciers form in the north, move south, and then retreat at the next warming period. Many such cycles of glacial advance and retreat have occurred. Geologists can read the tale of glacial advances and retreats in the rocks, soil, river valleys, and lakes.

At the last (or most recent) glacial maximum (LGM), about 21,000 years ago, a mile-thick glacier sat on top of where I currently live in Washington County, Minnesota. The glacier extended south as far as Nebraska. Having a huge ice sheet covering the northern part of our continent created massive changes to habitats that we know today. In short, plant communities were shifted south, which meant that the animals that live in these habitats were similarly affected. It is interesting to ponder where all of the plants and animals that currently cover North America went during glacial advances.

Surely, one might think, some species would have become very rare, because the southern United States covers far less land and habitats must have been greatly compressed in size. There must be some limits to how far south into Mexico and Central America northern species could have survived, given different soils and climates. But how could we begin to know where a species that we consider familiar today, like the Wild Turkey, might have survived glacial times. Some new scientific advances provide potential answers.

Environmental scientists have devised methods for predicting where particular species lived at various times in the past, such as the LGM and the last interglacial (LIG, 120,000 years ago). In short, you take the geographic coordinates of where the species occurs today and build an "ecological niche model" using various aspects of the environment, like temperature and rainfall. The model allows you to predict where the species occurs today, and since you already know this, you can see how well the model works. In most cases, it is very good.

Next you predict where these environmental conditions existed at the LGM using estimates of what the climate was like when the last glacier was at its maximum (southern) extent. This provides a prediction of the distribution of your species of interest at that time. The same procedure is used for the predicting distributions at the LIG. This method assumes that species lived under the same environmental conditions now, 15,000 years ago, and 130,000 years ago, but most people feel this is a reasonable assumption. I wondered, for example, where the Wild Turkey lived during glacial advances. Would Stone Age people have been able to hunt them?

To reconstruct Wild Turkey distributions, I went online to the breeding bird survey, which is supported by the U.S. Geological Survey's Patuxent Wildlife Research Center and the Canadian Wildlife Service's National Wildlife Research Centre. Each spring thousands of survey routes are run by volunteers, and their data are recorded online; the survey began in 1966 and is still going. I was able to download over two thousand point localities for Wild Turkey during the breeding season. Many of these localities are places where turkeys have been introduced, but at least they provide locations of current occurrences. To be more precise, I would probably cull records from places like Minnesota, where I personally don't think they occurred historically. Without human intervention in the form of agricultural crops and bird feeders, coupled

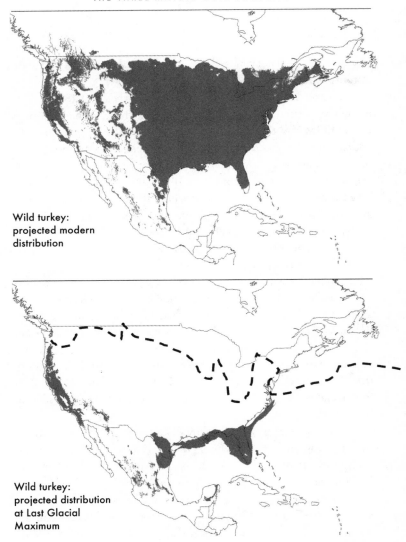

Wild turkey:
projected modern
distribution

Wild turkey:
projected distribution
at Last Glacial
Maximum

Top: Map of North America showing estimated distribution of Wild Turkey (dark gray) at the present time, derived from the ecological niche model. *Bottom:* Dark gray shows the estimated distribution of Wild Turkey at the Last Glacial Maximum, 21,000 years ago, assuming that turkeys then lived in places with the same climate characteristics as they do today. Note that areas currently offshore were above sea level at the Last Glacial Maximum. The dotted line represents the maximum southern extent of the glacier at the Last Glacial Maximum.

with numerous introductions, I doubt if Wild Turkeys could survive "typical" Minnesota winters by themselves. I'm aware that the Minnesota Department of Natural Resources doesn't agree and says they were native to Minnesota, but that's beside the point.

The model makes a good prediction about where turkeys should occur today, suggesting that it captured the main environmental features that determine where Wild Turkeys live. The analysis then tries to find where the conditions that the species exists in today occurred at the LGM. As you would expect, turkeys were displaced well south of the ice sheet (the dashed line on the map shows the approximate southern limit of the last glacier). For sure, no turkeys were anywhere near your current hunting spot in the Midwest.

You will also notice from the map that the projected distribution of Wild Turkey spills out of the current-day U.S. borders into what is today ocean, which might seem a failing of the method. However, when the last glacier was at its maximum, sea levels were lowered, and much more land was exposed around our borders; turkeys were predicted to have occurred there, as the map shows.

It looks like good places to find turkeys 21,000 years ago were south of the current southern border of Louisiana and east of the current eastern seaboard. Incidentally, if somewhat off topic, if you own low-lying coastal property in this current period of global warming, you might want to sell sooner than later.

Reconstructions of past distributions give us insight into where species like turkeys that today live very far north might have survived during periods of maximum glacial advance. One thing is sure, Wild Turkeys persisted in the face of major global climate change, at least over typical time scales of glaciation. Wild Turkeys were forced south, survived there during the LGM, and subsequently expanded their range northward to occupy their "present" range (whatever that is). This north-to-south-to-north cycle of range displacement likely happened many times during the past 2 million years. I think we can conclude that Wild Turkeys are survivors, which might explain why introductions have been so successful.

16

WOLVES, COYOTES, AND DEER

Perspective is a great help in identifying some animals in the wild. For example, if you see a canid (doglike animal), you wonder if it's a coyote or a wolf. If there's just one, it can be hard to judge size. Through the years I have heard of reports of lone wolves in my neck of the rural areas of the Twin Cities in midwinter, but my gut feeling is "not likely, but possible." I think that one would most likely see several wolves, and if they're around, there would likely be lots of reports. I realize there could be the lone "explorer." But my guess is that these observers saw a coyote, which can be confusing to identify when alone, at least to those of us not used to making a quick identification of a lone individual (I always forget that the wolf has the tail held out, coyote down). But my attention was caught by a recent scientific article that might shed some light on this issue, one of interest to deer hunters as well.

Coyotes are thought to have evolved as a prairie canid that catches primarily small prey. That coyotes in the northeastern United States are larger than coyotes elsewhere has been known for some time. We also know that northeastern coyotes eat a larger proportion of deer than those in the West and, not surprisingly then, do not avoid forested areas where deer hang out. The question is, why are coyotes in the Northeast so much larger than coyotes elsewhere? One possibility is that the environment in the Northeast favors increased size, so that if a western wolf were transplanted and grew up in New York, it would be big too. This is referred to as *phenotypic plasticity*. To further understand this, imagine you could clone a person and let one clone grow up in Minnesota and one in a third-world country. The clone in Minnesota would be bigger in spite of being genetically identical, simply because of better nutrition.

Another possibility, which was investigated by Roland Kays (New York State Museum in Albany) and colleagues in an article in *Biology Letters*, is that the "coyotes" in the Northeast are actually wolf-coyote hybrids, and as result of having genes from the bigger wolves, the "coyotes" are larger than their western counterparts and more likely to prey on deer. They used some standard molecular techniques to determine that a large fraction of northeastern coyotes carry wolf genes. Furthermore, they noted that the skulls of northeastern coyotes are not just larger, they are proportionately larger; in other words, if you took a picture of a western wolf and blew it up 20 percent, the skull would also be 20 percent larger. But in northeastern coyotes, it's like the skull is 30 percent larger when the rest of the body is only 20 percent larger (these are not exact numbers but for illustration). Kays speculated that the wolflike skulls of northeastern coyotes were "associated with strong bite forces and resistance to the mechanical stresses imposed by large, struggling prey." Translation: holding on to not-yet-dead deer.

How did this happen? Reliable records show that western coyotes "invaded" the eastern forests starting in the early 1900s using forest openings created by agriculture. They took a two-pronged approach: some went north around the Great Lakes, where they encountered "Great Lakes wolves," and some took a southern route, through Ohio, where wolves had already been extirpated (driven out). Although coyotes and wolves are supposed to be separate species, meaning they don't hybridize, many coyotes taking the northern route hybridized with Great Lakes wolves. Today, the pure wolves are gone, and in their place is a population that is a hybrid mixture of wolves and coyotes. So, wolves are gone, but they left behind a genetic legacy—a bigger coyote that eats more deer than its western ancestors.

What about the "wolves" seen by my friends near the Twin Cities? Could they be large-bodied, coyote-wolf hybrids? More wolves live in northern Minnesota than in the recent past, and they were removed from the endangered species list; in fact, a hunting season has been enacted. Possibly wolves are actively hybridizing with coyotes, making some "uber-coyotes" like those in the northeastern United States. This is a question for genetic analysis, which I hope is or will soon be under way in Minnesota. Incidentally, genetic analyses have provided little evidence for domestic dog and coyote hybrids.

Should deer hunters be concerned? I doubt it. Here is a March 3, 2009, press release from the New York State Department of Environmental Conservation: "Hunters harvested approximately 223,000 deer in the 2008 season, a 2 percent increase over the previous season." In Pennsylvania, hunters averaged 410,256 per year from 1998 to 2009, and the 2008–9 harvest exceeded the previous year. These are areas where there are big-bodied coyotes, so no, they apparently have not succeeded in reducing the herd all that much. We should remember that wolves were a part of the natural ecosystem of much of our country. In their absence, many ecological changes have occurred, such as the expansion of the deer herd (whether it's cause and effect is not clear).

In the absence of wolves, coyotes expanded and now appear to be hybridizing with newly reintroduced wolves. This is an example of an experiment that is human induced, but it is not unnatural. Climate-induced habitat changes have always occurred and allowed different species to expand their ranges. In these cases and the present one, we'll just have to wait and see what actually plays out. In the meantime, I don't see any reason to change the way we manage wolves or coyotes. Plus, I noticed that "wolves" in Ontario also carry coyote genes—so perhaps "wolves" will be getting smaller as a result of hybridizing with coyotes!

17

LEAD, LEAD, EVERYWHERE?

What is a common denominator in the following human physiological problems?

- impaired motor function
- impaired cognitive ability
- reduced intellectual development
- impaired kidney function
- reduced endocrine function
- impaired tissue growth
- impaired reproductive development
- spontaneous abortion
- decreased brain volume
- behavioral abnormalities

No, they are not characteristics of your in-laws, as relevant as the last two might seem. If you answered lead in our bodies, you would be right (or you read the title of the chapter). This laundry list of human maladies caused by lead makes clear why we are now so aware of getting the lead out of our environment. Not to mention not inhaling it at the gas pumps anymore. It's really toxic stuff, and it's not like some other things in our environment that are beneficial in small amounts.

As far as we know, lead has no biological role. However, a handy online calculator tells us that the average two-hundred-pound person has approximately 0.15 grams (0.005 ounces) of lead in their bodies, so we ingest or inhale it nonetheless. And remember earlier writers' assertions that the form of lead in bullets would prevent human contamination? Although it is true to a degree, lead bullets that have been fired from a gun likely do provide a source of lead contamination. Of course, given we absorb lead from other sources, it is difficult to determine whether such exposure by itself would lead to levels that exceed the Centers for

Disease Control exposure level threshold. But no matter, I think I'd prefer not to add any new sources of lead contamination to my tissues and bones. Age has caused enough problems.

Lead products break down over time, and lead gets into the soil, water, and plants. Once it gets into the bloodstream, it builds up in soft tissue and comes to rest in bone. Once in bone, it can remain there for decades, where it occupies molecular positions normally occupied by calcium. The reason for this is that the chemical properties of lead mimic those of calcium. Because calcium plays a huge role in the working of our nervous system and lead does not work the same as calcium in nerve functioning, lead poisoning leads to serious neurological effects.

A 2012 article by W. Grainger Hunt in the *Journal of Raptor Research* reviewed the toxic properties of lead in general and how they affect birds of prey. I found it a sobering lesson on the long-term effects of how we have treated our environment and can only say I'm sorry for my part in adding lead sinkers and shot to the environment. If I'd have known then what I do now, I would have acted differently. However, I recently tried to buy some nontoxic sinkers at the private bait shop on Lake Itasca in Itasca State Park only to be told that they didn't stock them because it was legal to use lead—on a lake to which many come to see nesting loons!

In the article, Hunt reported on the effects of lead ingestion on birds of prey. We have known for a long time that birds feeding on waterfowl ingested lead pellets, which were not lethal to the ducks (at least at the time they were shot) but built up in birds of prey that scavenged them. This observation led to bans on lead shot in waterfowl hunting. Despite much angst, we seem to have accepted this imposed sanction on lead shot, no doubt for the better.

More recently, researchers discovered that gut piles often contain lead fragments from bullets. In one study, 90 percent of gut piles contained lead fragments, and 50 percent contained one hundred or more lead fragments. Scientists think that the reintroduced population of California condors in Arizona will not be self-sustaining unless the birds are monitored and treated for lead poisoning, which they get by feeding on gut piles from hunter-shot elk and deer. Many lead-poisoned condors develop crop stasis (where food doesn't pass through the digestive system), and they end up on the ground unable to fly and then starve, usually undetected.

Bryan Bedrosian and colleagues conducted a study on lead in Bald Eagles, published in the journal *PLOS ONE* in 2012. They worked in the Jackson Hole valley of northwestern Wyoming from 2005 to 2010 in areas where exposure to lead is limited because there is no varmint, predator, waterfowl, or upland game hunting, and no live bait is allowed for fishing (which should limit exposure from lead sinkers). As found in the other study, the birds get lead from gut piles, not from waterfowl.

They captured eagles at different times of the year and took blood samples to see if lead levels in blood were higher during the hunting season. This would implicate gut piles as a source of lead. They tested eighty-one blood samples before, during, and after the big-game hunting seasons and found that lead levels were significantly higher during the hunt. In particular, they found that 24 percent of eagles (irrespective of sex or age) had clinically elevated lead levels during the hunt, whereas none did in the other two seasons of the year. However, 93 percent of all sixty-eight nonnestling eagles tested had been exposed to lead at some point.

The researchers did driving surveys and fitted some birds with radio transmitters. They found that the number of eagles increased during the hunting season and that the birds returned in subsequent years. This means that the eagles recognize this fall food source and return to take advantage of it, albeit at risk of increased lead exposure.

The researchers attempted to influence the incidence of lead in eagles by providing a free or reduced-cost box of nontoxic shells (well, I'm sure their grants paid for them, so "free" is relative) to hunters in Grand Teton National Park and the National Elk Refuge. The program was successful, and they concluded that "the use of non-lead ammunition significantly reduced lead exposure in eagles, suggesting this is a viable solution." This is good news. California has recently instituted a ban on lead bullets, but compliance is estimated to be 80 to 95 percent. As good as this is, clearly we need more evidence to convince the public that lead bullets are bad for the environment. The Bedrosian study now provides that evidence. It certainly supports the growing sentiment for outlawing or severely curtailing the use of lead bullets.

And I would like to put in a plug for my favorite local hunting club, Wild Wings of Oneka, which has required steel shot for sporting clays and hunting for many years. Some clubs host national sporting clays championships where competitors shoot over marshlands. As many as one thousand shooters discharge four hundred or more rounds in a few days. That amounts to around twenty-five thousand pounds of lead. Multiply that by decades . . .

We obviously did not purposefully contaminate our environment with lead, and now that we know the extent of its toxic effects, we are more careful. It does make you wonder, though, about what we are currently adding to the environment that might someday turn out to be the next lead threat.

POLITICS AND THE LEAD AMMO DEBATE

One of my essays on lead ingestion by Bald Eagles generated some controversy after it was published. One reader wrote that my article contained misinformation and was politically motivated. I had to chuckle at the thought of my having a political agenda, but given the growing importance of the lead debate, further discussion is bound to occur.

What is the political agenda? Well, for example, Safari Club International has joined the National Rifle Association (NRA) in opposing bans on "traditional ammunition," which of course means lead. The reason? The NRA has decided that the movement to oppose the use of lead bullets is actually an attack on our Second Amendment rights. Yes, that's a political agenda.

My "political" agenda is this: eight shotguns, two rifles, two bows, two handguns, three pointing dogs, and my permit to carry. I do not work for or represent any group that has a dog in this fight, except, perhaps, my actual dogs.

I personally think it challenges credibility to suggest that opposing lead bullets is opposing the Second Amendment to the U.S. Constitution. I admit to doing a double take when I first read this stance. Didn't we just go through this with nontoxic shot in waterfowl hunting? Did we lose our shotguns or our right to hunt ducks and geese? My family enjoys eating the local geese in our freezer and a nice bunch of mallards from Manitoba. And yes, we used nontoxic shot.

I can only surmise that the real agenda is the fear that many will not be able to afford nontoxic bullets and therefore will give up hunting, which would be bad. Yes, I wince at the price of a box of nontoxic shotgun shells or copper bullets. However, many prices have come down as demand has gone up. The price of ammo is not the big driver in the costs of hunting. Instead of spreading fear, the NRA should lobby for cheaper nontoxic bullets. These bullets do the job, as my son and I discovered in a recent hog-hunting trip to south Texas.

Also, there are two issues, which should perhaps be considered independently: the type of bullets used for hunting, and the type used for target practice. Should we stop using lead bullets for both? We could use lead bullets for target practice only, but lead and copper bullets can have different ballistics, and so you'd have to sight in your rifle for whatever nontoxic bullet you might hunt with.

Perhaps we could limit practicing with lead bullets to particular well-marked areas, from where maybe in time the lead could be recovered? But maybe it's irrelevant—random lead bullets might not have a demonstrable effect on the environment in general for thousands or more years. We would need to do some scientific surveys, and perhaps they would find trivial lead levels in our favorite deer woods as a result of lead rifle bullets. I still feel that adding lead to the environment without any possible removal is a problem worth considering.

The letter went on to note that I failed to mention that the eagles that had ingested lead from carcasses (this is not in dispute) had sublethal levels of lead in their blood. However, this finding raises a couple of concerns. First, lead can be deposited in bone and build up over time, to be released later at potentially toxic levels. Has it happened yet with eagles? Apparently not, but lead has been shown to be toxic to waterfowl and California condors.

The second concern is whether we think it is OK for eagles to have lead in their blood as long as it is at sublethal levels. Is the same true for you or your kids? If someone offered you venison with sublethal levels of lead in it, would you knowingly eat it? Recall that lead has no known biological function and is only a toxin. You won't die from sublethal levels, but the long-term effects are almost certainly not good. Maybe it's better not to roll the dice? The North Dakota Department of Health noted: "In young children, lead exposure can cause lower IQs, learning disabilities, stunted growth, kidney damage, attention deficit disorder (ADD) and attention deficit hyperactivity disorder (ADHD)."

Lastly, the letter commented that "Minnesota bald eagles have been so successful that the DNR no longer even counts them and is recommending they be taken off the list of species of special concern. I guess these details did not fit with the writer's narrative." Minnesota Bald Eagles have indeed rebounded, as they have in many places. Apparently, then, lead must not be a concern for our local eagles, and only eagles in the western United States ingest lead?

I consulted Pat Redig from the University of Minnesota's Raptor Center, who is a veterinarian specializing in medical problems with raptors (birds of prey, like eagles). Contrary to the letter writer's optimistic view of Minnesota's Bald Eagles, Redig wrote, "We've had sixteen bald eagles poisoned with lead in the past five months—most fatal." One can read about lead poisoning on the Raptor Center's web page.

Much was also made in the letter of a study of potential lead ingestion by hunters in North Dakota. The North Dakota Department of Health tested the blood of 738 hunters who had eaten venison harvested with lead bullets. The letter stated "that not one individual had lead levels considered elevated. There is no link between increased blood-lead levels and consuming hunter-harvested meat." I looked up the study online.

The North Dakota Department of Health web page summarizing the results of the tests of North Dakota hunters began with this statement: "The study shows a link between eating wild game shot with lead bullets and higher blood lead levels." Furthermore, the report stated, "In the study, people who ate a lot of wild game tended to have higher lead levels than those who ate little or none. The study also showed that the more recent the consumption of wild game harvested with lead bullets, the higher the level of lead in the blood." I can only assume that the reader misread the study, but his summary defines misinformation!

Now granted, there are many ways to ingest lead, and maybe people who ate more game had higher levels because they used to be painters or worked with leaded gasoline. However, a well-known bullet manufacturer stated that their bullets had 60 to 70 percent retention, meaning that 30 to 40 percent of the lead was shed via fine particles into the wound channel. Although levels might be relatively low, I'd prefer to avoid any extra lead in my blood (or that of my family). The North Dakota Department of Health, perhaps erring on the side of caution, posted these recommendations based on the blood lead level study:

- Pregnant women and children younger than six should not eat any venison harvested with lead bullets.

- Older children and other adults should take steps to minimize their potential exposure to lead, and use their judgment about consuming game that was taken using lead-based ammunition.

- The most certain way of avoiding lead bullet fragments in wild game is to hunt with non-lead bullets.

The last recommendation is consistent with a statement in the article by Bedrosian, which I summarized: "Unlike many environmental problems there is a straightforward and easy solution to toxic lead exposure in wildlife from rifle ammunition: nonlead alternative ammunition." In sum, I think that limiting the lead in our systems and the environment is a good idea. I don't think it will result in my losing my Second Amendment rights. Ammunition manufacturers should develop less expensive alternatives to lead bullets than the current nontoxic "premium" bullets. Until then, I think I'll bite the nonlead bullet.

19

GETTING THE LEAD OUT (OF CHUKARS)

At a recent meeting of ornithologists in Jacksonville, Florida, a paper by graduate student Justin Bingham caught my eye. He presented two papers on lead ingestion by Chukars, and since I regularly eat Chukars from a hunting club, I was immediately interested. Plus, it's hard not to be aware of the negative effects of lead in our environment.

His first talk, titled "Causes of Lead-Pellet Ingestion by Captive Chukars," addressed the question, why would a Chukar be dumb enough to eat lead pellets? I thought about some of the things my dogs eat and figured I'd hear him out.

First, some background. Bingham and his colleagues found that around 10 percent of feral Chukars in their study area in Utah had at least one lead pellet in their gizzard, and a bit higher percentage had relatively high concentrations of lead in their livers. It is known that a single #6 lead shot can kill a bird the size of a Chukar if it stays in the gizzard. If it isn't quickly passed through the digestive system, the grinding action of a Chukar's acidic gizzard results in the bird getting a pretty good dose of lead in its bloodstream. Unlike some things that are actually OK in small amounts in the body, lead probably does not occur in animals naturally, and when it does get ingested, it serves no useful purpose.

One might think that a Chukar would recognize a lead pellet for what it is and just avoid it. Or maybe they think the pellets are rocks and ingest them for grit to help the gizzard grind food. The wild Chukars Bingham and his colleagues studied prefer seeds from Indian ricegrass, which produces a nutritious seed, one that was a staple food in the diet of Native Americans. These seeds are dark colored and about the size of a #6 lead pellet. Bingham experimented with captive Chukars and found that they often mistake pellets for seeds. In a clever experiment, he showed that when different, light-colored seed was mixed together with dark-colored lead pellets, the Chukars were much better at avoiding pellets. He was able to show that they are simply confused by the pellets and not just ingesting them as grit.

The toxic effect is heightened if the ingested pellets are "weathered" rather than relatively new. In his experiments, Bingham made pellets weathered by combining them with some coarse materials and tumbling them around. This result is even worse news for Chukars (and other things) because pellets can probably remain in the environment for three hundred or more years, and they weather with age. And we have to assume that his method for making pellets weathered replicates natural processes that affect lead pellets on the ground—but I thought it was a clever idea.

An interesting twist on the story comes from another item in the diet of Chukars, namely allium, or wild onion. Wild Chukars eat quite a bit of wild onion, and Bingham wondered why. In his experiments, he fed Chukars wild onion (and garlic!) and found that they suffered fewer effects from lead ingestion, suggesting that Chukars might self-medicate themselves by eating wild onions! Garlic seemed to work too. Apparently compounds in onions inhibit or reduce the uptake of lead.

The second talk covered physiological effects of lead ingestion by Chukars. Short story: it isn't pretty. The lead that leaches into the bloodstream poisons various organs and leads to a drastic decrease in muscle mass. If there's a good side to this, lead concentrations do not get very high in breast muscle, so if you were to eat a Chukar in the early stages of lead poisoning, it might not affect you. But the breast muscle mass can be greatly reduced. Also, some scientific studies suggest that lead does accumulate in muscle tissues and can be passed to humans, which is not healthy.

I've hunted wild Chukars once, and although I think I'm a good shot, I was humbled by these birds. They are nothing like Chukars on game farms, as they are wary and extremely fast. Those characteristics coupled with the fact that they live in some pretty steep places make shooting them a real challenge. I think I was one for nine, and I shot several times at some birds. Obviously this is nothing to brag about. The point is that if my experience is typical, quite a bit of lead is out there in the environment.

There have been many studies of the concentration of lead pellets in the environment. In an area of New Mexico, authorities estimated that 340,000 doves were harvested in 1985 in a five-county area and that on average hunters fired eight shells per dove. If each shell had one ounce of lead, fifty-plus tons of lead were deposited. In Minnesota, Richard Kimmel and Molly Tranel reported that the Minnesota Pollution Control Agency estimated that 2,610,720 pounds of lead shot were used annually in Minnesota in hunting and at shooting ranges. So, the potential to ingest pellets might actually be higher than we think.

Lead pellets can be highly concentrated near water sources, which are not that common in the western United States. Hunters shoot a lot of shells at Chukars and doves that have come in to drink at stock tanks. Birds and hunters return over and over to these places, leading to the concentration of pellets near water sources.

We've been keenly aware of bans on lead shot owing to its toxicity, especially near water where hunters over decades have discharged their shotguns repeatedly. We now ban the use of lead shotgun shells when hunting waterfowl and on many government-owned lands. However, in more upland areas, that's often not the case. I and thousands of others have hunted Ruffed Grouse with lead shot, and I think I'm on safe ground saying that way more shots are taken than the number of grouse that make it to the table. So I've always wondered about the probability of a Ruffed Grouse even finding a lead shot, no less eating it. The area of Utah where Bingham works is dry and not heavily vegetated, so there is a low accumulation of plant material after the growing season. This keeps shot closer to the surface, if not directly on top of it, and it weathers.

In our Ruffed Grouse woods, leaf litter builds up on the forest floor as leaves are dropped each autumn, which might cover pellets in a few years. However, they can be scratched up by birds like grouse searching for food. And if a single pellet could be fatal, we should be concerned. Or should we? A scientific paper published in 2005 reported that in Quebec Ruffed Grouse had very low levels of lead (and none was found in Rock Ptarmigan and Willow Ptarmigan). A survey in southwestern Virginia of sixteen Ruffed Grouse found no contamination. I've never looked closely at the grouse I've been lucky enough to get, but maybe hunters should start to examine the crops and gizzards of the Ruffed Grouse they harvest.

The tons of lead going into the environment over the years can leach into the soil and groundwater and have some far-reaching effects on both wild animals and people. I've never worried about using lead in upland hunting, but I've decided to go nontoxic in the uplands as well as over water. Yes, a bit more expensive, but it's probably time we made the switch.

20

SOUNDING THE ALARM, MOURNING DOVE STYLE

One fine spring day I sat in a ground blind with my bow, waiting for Mr. Tom. This usually involves lots of inactivity, and to pass the time, I watched a Mourning Dove gathering nesting material from the ground. Mourning Doves make a nest consisting of a loosely woven bunch of sticks and grasses—typically it is so loosely put together that you can see through it from below. They lay two eggs. They incubate a precise number of days, and if the eggs don't hatch on time, they abandon the nest. Sensible indeed.

Mourning Doves are among the top ten most common birds in North America, and more are killed by hunters than any other game species. Large numbers are harvested in southern states, and some old friends of mine in Louisiana go out after work in the autumn to abandoned fields and collect a dinner's worth. And they are worth it, they are great eating.

The dove I was watching would fly down, pick up some grass or twigs, and fly off, making the characteristic wing whistling of the species. American Woodcock make a similar sound with their wings. I was through a few cups of coffee while watching this Mourning Dove, in lieu of any sign of Mr. Tom, and it dawned on me. What, if anything, is the function of the wing whistling? I've heard it for so long it is just part of watching Mourning Doves. But could it have a purpose? Did anyone know?

I did a quick Google search (yes, even ornithology professors use this to track down things they are supposed to know). Although I had a faint hope that perhaps no one else had thought of this question, I was not surprised, and only mildly disappointed, to learn that I was well behind the knowledge front! It's really hard to think of things no one else has thought of.

Dove wing noises are in fact well known and described. The

noise comes from air passing over the wing feathers. However, I am unsure if the dove's wings have any special modifications, as those of the woodcock do, where the outer primaries are modified for making sound. A check of specimens in the collections at the University of Minnesota's Bell Museum of Natural History showed that the outer part of the primaries seems to have greater surface area than those of other birds, but the role of this in making sound is not known. There have, however, been experiments aimed at discovering a possible function of the wing noises.

Robert Magrath of the Australian National University in Canberra did an experiment with Crested Pigeons. He had noticed that birds taking off under normal circumstances had different wing sounds than those taking off when frightened by a predator; in the latter case, the bird takes off faster and at a steeper angle, which produces a different wing sound. Magrath and a student recorded wing sounds from birds taking off normally and when startled. Knowing they might have to wait a long time to record birds taking off when startled by a predator, the researchers had a clever idea. They "flew" a glider shaped like a hawk over feeding stations where the pigeons were eating. When analyzing their recordings, they could see that the wing whistle differed between birds taking wing under the two different conditions. Then, they played the recordings to flocks of Crested Pigeons eating at a food tray. None of the fifteen flocks responded to the normal wing whistle—they just stayed and kept eating. But when flocks heard the wing whistle of startled birds, eleven of the fifteen flocks flew off. So the startle wing whistles are probably a way of sounding the alarm.

Magrath was not the only one to test this idea experimentally. Seth Coleman from Texas A&M University did a similar experiment with Mourning Doves. He got recordings of normal wing whistles and startle wing whistles, the latter by releasing a house cat when doves were at his feeder (this might be one of the only valid reasons for letting cats outdoors!). He then noted the responses of feeding doves to the recorded sounds of unstressed

Mourning Doves taking off and those leaving in a hurry because of a "predator."

Coleman's results were consistent with Magrath's. Mourning Doves were more likely to leave and to stay away longer when they heard the startle wing whistles as opposed to the wing whistles of a Mourning Dove lifting off normally. The same was true with other species, like the familiar Northern Cardinal, that shared the feeder with the doves.

Most bird species have vocal alarm calls. Some are intended to sound the alarm without revealing the precise location of the sender, and other calls are intended to allow other birds to assemble in one place to help drive off a predator. The latter sounds are easy to locate, the former harder to locate. It is pretty clear that Mourning (and other) Doves use a nonvocal sound to signal a threat to other doves (and other species have homed in on their emergency frequency). Actually, it's pretty useful—the sound a dove's wings make when it is fleeing danger can warn others, or at least close kin.

We don't know whether doves can regulate this sound, or whether it is essentially an incidental by-product of the way their wings are shaped; that is, maybe the dove isn't interested in saving anyone but itself. But if you're a dove and you hear that another dove is taking off in a hurry, there is probably a good reason to do so yourself. Pronto.

As I was thinking about wing noises from Mourning Doves, I heard a vocal warning from a bird! Two jakes had come up on my decoy and for some reason were concerned. One gave some warning clucks that jolted me back to the task at hand. The other jake stood in the clear at twenty-nine yards. He should have paid more attention to his friend's warnings.

21

MOANING MOOSE AND TOPI LIES

I would guess that in general what we know about the behavior of animals is loosely related to their size. Small animals are often hard to observe and study. Conversely, you might think that we probably don't have much left to learn about the behavior of an animal as large as a moose. I mean, the problem isn't that they are hard to see. A recent scientific paper proved that wrong.

But before I review its findings, a little Darwin. He recognized that there were some common aspects to the differences between males and females among species. For example, males are usually larger than females, have brighter colors, and often possess spectacular "ornaments," like the antlers of deer and moose. I'm sure that anyone who has looked at a giant moose mount has wondered what it must be like to have that thing on top of your head. Getting through brush must be a real pain. And then, if you dwell on it and recall that moose antlers can weigh up to sixty pounds, you might think bull moose should lie down a lot and give their necks a rest. I haven't seen any signs for Moose Chiropractic Clinics, so they apparently endure, and those racks must be worth it.

Darwin figured out why moose have such gigantic antlers, as opposed to a much smaller version that would more simply let everyone know that it's a boy and not a girl moose. The biological process he figured out is called sexual selection. Basically, if a male moose has larger antlers than others and these allow him to be dominant over other males, he wins the jackpot (in evolutionary terms) because females choose the dominant male with whom to mate. In fact, he may prevent most other males from mating at all, resulting in most of the young being sired by himself and inheriting his impressive antler size. This leads to successive generations of moose having larger and larger antlers, resulting in some truly impressive racks. This doesn't mean that all moose have equally big antlers; rather they have a distribution of sizes, and the bigger males are more successful. And there has to be an upper limit to antler size, which we might have reached, but we won't know for a few thousand years.

Through all of this, options available to females are often overlooked. A female moose weighs a lot less than a bull, even a relatively young bull, and perhaps she could be forced into mating with a male that was not her first choice. Terry Bowyer from Idaho State and colleagues published a paper in the journal *Behavioral Biology and Sociobiology* describing how they followed radio-collared moose in Alaska for several years and watched behaviors of cows and bulls during the rut. They focused on moose vocalizations.

Although I've not hunted moose, I've seen lots of videos where the hunter or guide gives a grunt call that seems pretty easy to imitate. Many of the scenes depict the guide with a canoe paddle over his head, pretending to be a rival bull—eyesight is not a moose strongpoint, but they can hear. Male moose have two types of calls, a grunt given to females they are courting and a modified version used when they are wandering around looking for mates. I thought that these were what Bowyer and colleagues were studying, but instead they were interested in vocalizations given by females termed "protest moans," described as "plaintive, undulating calls of 3–5 s duration." These moans have an interesting purpose.

Consider a female moose in the breeding season that comes into heat. This is a relatively short window of time, and from the female's perspective, she wants to mate with the best male she can so that her offspring will have the best genes they can get. All of sudden, she's confronted by an amorous male, intent on being her partner while the dominant bull is preoccupied elsewhere. But he's not good enough. He's a smaller bull, and although he might someday be the dominant bull, today he's not worth the risk. "Risk" because cows have relatively few breeding efforts over their lifetimes, and each one is costly in terms of time and energy contributed by the female. Picking the wrong partner has negative consequences for a cow moose. This gets at a basic rule of sexual selection: males should mate as often as possible; females should be choosy. Thus, a few males do the lion's share of the matings, whereas most if not all of the females are bred. Still, they have to set high standards.

But this smaller, nonpreferred male is still a lot larger than she is, and he is clearly interested. Her solution? A white knight on a horse? Perhaps something more subtle. Bowyer and colleagues figured out the cow strategy by watching 105 Alaskan adult male moose, which they ranked as small, medium, and large, and how they responded to protest moans given by cows. They found that a cow being "hassled" by a lesser bull gives a greater number of protest moans. In fact, relative to when being courted by large males, protest moans increased by three times when being courted by a medium-sized male, and four times when courted by a small male.

Frequent protest moans quickly bring the top male to the scene, which is not good for a small bull. In fact, male-male fights were twice as likely right after the protest moans, and the little guy loses. So, the cow's strategy is to let the harem master know one of the satellite bulls needs to be put in his place, and he's more than obliging.

Interestingly, though, cows also give some protest moans when being courted by a harem master. This seemed a bit odd to me, as if he is the preferred male, why risk attracting other males and creating a scene? The researchers suggested that indeed this is the point. It sometimes results in a challenge to the harem bull, who has to prove, once again, that he is the alpha dude! Only the best for our cow, and she's not above calling the question!

Lest we feel sorry for the underdogs who get ratted out, elsewhere it works both ways. In the topi, a grassland antelope from Africa, females and males congregate during the breeding season. Females visit several males and will mate with multiple males. This is also a somewhat dangerous time, as such congregations attract lions, who are hoping to take advantage of a distracted topi. Topi give an alarm call when a large predator is sighted. Male topi, however, are not all about truth in advertising. A male about to have a female walk away will give a false alarm call, with no lions nearby, attempting to trick her into staying and increasing his mating chances! That's right, male topi lie. I guess we have a gender standoff in the signaling department. Cow moose and male topi both use vocalizations to achieve their goals.

22

TURKEYS AND LOVE
WHAT'S ACTUALLY HAPPENING OUT THERE IN SPRING?

The spring mating season of the Wild Turkey signals a welcome change from Minnesota winter. As hunters know, solitary toms or groups of toms patrol for available hens, each of which has figured out where they are going to nest. We know that this is the time when turkeys mate, but what are the "rules" or traditions? For example, why do toms hang around together if they are competing for mates? Why not space out and defend territories, like robins? Do all toms eventually mate equally during the breeding season? Does one tom father most of the offspring in an area, like prairie chickens or Sharp-tailed Grouse? Are hens really choosy about which toms they mate with? Do hens fall in love with only one tom? Or when the love of their life is preoccupied, do they "stray"? Can the poults answer the question, "Who's your daddy?"

Until recently, we didn't know. A researcher from the University of California, Berkeley, Alan Krakauer, used some sophisticated DNA tests to figure out which turkey did what and with whom. These tests use blood and tissue samples from adults, young, and eggs to reveal whether all the eggs in a nest are from one male and one female, or whether turkeys are "unfaithful." The results of the tests are very compelling and would stand in a court of law as flawless indicators of paternity and maternity!

Krakauer studied Wild Turkeys in California, and his DNA testing provides the first glimpse of what actually happens out there in the spring when turkeys are chasing each other . . . and we're chasing them. First, about half of the nests could be attributed to one female mating with one male, a system we call monogamy. So then, Wild Turkeys have some "morals" after all. But about 25 percent of nests contained eggs from two hens (usually one or two eggs from one female, the other six to ten from the incubating hen). That means that in some nests hens were incubating eggs that were not their own! In several of these nests, the eggs dumped there by the nonincubating hen were fathered by the same male as the eggs laid by the incubating hen. One enterprising hen laid a couple eggs in another hen's nest and then laid a clutch of her own that she incubated—probably the best of the not-putting-your-eggs-in-one-basket strategy.

Now what about the rest of the flock? About 25 percent of the nests showed multiple paternity—that is, turkey hanky-panky. In these nests, the eggs laid by the incubating hen were fathered by two males (one nest had three separate fathers "indicted" by the DNA evidence). About 12 percent of all baby turkeys were from nests with eggs fathered by two males. So, having a half brother or half sister is pretty common among Wild Turkeys. This detail adds a little more perspective to what we're seeing out there in the spring. But it's not the end of the story, because what happens to males and females is different.

All females likely mate during the season. Not so for toms. Krakauer concluded that 60 to 80 percent of toms did not mate in a given year, although he indicated this could be an overestimate because he did not find all nests each year. Of this group of unlucky toms, about 25 percent were subordinate members of "coalitions"—groups of toms in which one is the "boss tom" and the others assist him by displaying to hens. Apparently having several displaying toms gets the hens more excited, and they mate with the boss bird (who prevents the other toms from mating). So, that's why there are groups of toms!

Obviously, one has to ask why in the world a tom would help another tom to attract the attention of hens. The answer is that they probably "hope" to inherit the dominant role in the future, if and when something bad happens to the boss (which admittedly, we hunters are out there trying to make happen!). That leaves the other 35 to 50 percent of males who were just losers. So, a lot of toms go to a lot of effort that doesn't pay off in the love department, but they hope that the boss meets an early demise.

Just how successful are the successful toms? Krakauer found eggs from a single tom in up to four nests, but no one tom dominated the breeding area, as is common among prairie grouse. So not only do some toms not mate, but those that do mate may have multiple partners. Males with a minority share in a nest did sometimes father young in other nests as well. Krakauer observed in the field that these color-marked males were never subordinate members of coalitions who snuck away from the boss but were dominant males in other coalitions or solitary displayers.

So, the love life of the turkey is a mix of goings-on. About a quarter or more of all females take two or more partners for at least a few of the eggs. We aren't clear why she does this, or even why she would mate multiple times with the same tom, because it is known that domestic hens can store enough sperm from a single mating to fertilize an entire clutch. Having multiple partners can increase the genetic diversity of the hen's young, which could lead to more surviving offspring. This is a common "strategy" among birds. But there's a different twist in the Wild Turkey. Maybe Wild Turkeys routinely mate many times per clutch and would normally mate with the same tom (as in 50 percent of the nests sampled).

Perhaps later in the laying cycle many "original toms" had become too curious about some weird decoys and odd-sounding calling, and the hen took another mate by necessity. My guess is that hunting has little to do with turkeys choosing multiples mates (the Krakauer study was in a no-hunting zone), and I'd bet it's the norm whether there is hunting or not. But until the recent DNA evidence became available, we could say little about the consequences of the mating rituals of Wild Turkeys. Now when we see those broods in summer, we'll know the young turkeys might not be too sure who was their daddy or their mommy.

23

LOOKING BACK AT TURKEY SEASON
WHAT YOU MIGHT NOT HAVE SEEN

A quick glance is usually all that is needed to tell whether the turkey approaching your decoy is a mature tom, a jake, or a hen. Obviously, tom turkeys are larger than females and jakes, and toms are the most brightly colored, resulting from iridescent feathers and vividly colored patches of skin. What exactly *does* a hen see in a tom, anyway?

Since Charles Darwin's time we have known that females choose males that somehow indicate their superior features. What sort of features? Well, first of all, to be gaudy in the extreme, like a tom turkey, means you are visible not just to hens but to predators as well. So, simply being alive may mean you possess superior characteristics that have helped you survive in spite of your outrageous appearance and behaviors. This is sometimes referred to as the "handicap principle." But what if there are multiple toms to choose from? Do females choose males based on their beard length, feather brightness, how well they gobble, body size, or spur length, things we tend to emphasize? Or are there characteristics of toms that only another turkey can see?

Here, we should note that birds and humans see colors very differently. First, birds see an expanded color spectrum, because their eyes have our set of cones (color sensors) plus a fourth set. Second, birds see in the UV spectrum, whereas humans do not. So, when we look at turkeys, we may not see what it is that piques the interest of a hen in a particular tom.

You might ask, so what if toms differ in their true coloration and UV reflectance in ways we cannot see? Are hens making choices about which tom to mate with out of vanity? Maybe not. Turkeys are often infected with a protozoan parasite, and males with these infections can have altered feather colors. What if the more parasites a turkey has, the less bright or iridescent its plumage will appear? If true, females could choose males with "better" feathers to father their young because they have the fewest parasites, which would be a decided advantage for the offspring. In other words, if some toms have more genetic resistance to parasites than others, turkeys might see it in their feathers. And we, with our vision, would be none the wiser. At least that's the theory.

Geoff Hill and his colleagues at Auburn University decided to investigate this question experimentally using three groups of captive-reared young male Wild Turkeys. One group was infected with a single strain of a common turkey protozoan parasite, the second group was infected with a mixture of different species of protozoans, and the third group was kept free of infection. Their idea was to compare the newly molted feathers of males and see if having a parasite infection affected feathers. They found that exposure to the protozoan parasites affected certain aspects of the iridescent coloration, namely, brightness and ultraviolet reflectance. A male with a higher parasite load might look a little different to you and me, but it probably looks very different to another turkey. So, unbeknownst to us, turkeys can check each other out and determine who is more infected than whom! Now, the next step is to do the critical experiments to see if females actually do prefer males with lower parasite loads and brighter plumage.

But it won't necessarily be that easy. As we try to remember in scientific studies, "correlation doesn't prove causation." In fact, in another study, Richard Buchholz from the University of Mississippi found that parasite load was also related to length of the frontal wattles, or "snood" (that weird fleshy thing that hangs down from the head): more parasites, shorter snood, and another potential way for females to judge a male's quality. We know that males with longer snoods are chosen as mates more often by females and avoided more often by other toms (the mine-is-bigger argument seems to work here). So, maybe parasites have an effect on lots of things that influence which tom or toms females choose to mate with, and how toms decide among themselves who is the boss tom. In any event, it is likely way more complicated than we can determine with our relatively weak powers of vision.

These infections are not extraordinary, and many turkeys have varying levels of them. Choosing mates based on feather characteristics also helps explain why the sexes look so different. Females have an advantage in remaining drab and camouflaged because they alone incubate the eggs. Males with lower levels of infection are chosen over others, and if they can demonstrate their lack of infection by becoming more and more gaudy, "gaudiness" itself will increase in the population (because the others don't breed). In the process, male and female appearances come to differ more and more, and the current difference between the sexes can be thought of as a stage in a process. This process, termed sexual selection, likely has an endpoint at which extravagant male plumages and behaviors are costlier than the benefits they reap from them.

WHEN BLACK BEARS ATTACK!

On September 28, 2002, a father and son were bow hunting for elk in Idaho. The son was waiting in ambush in a clearing, while the father was a hundred yards away, calling. All of a sudden, two black bear cubs entered the clearing, followed by their mother. The sow charged the son, who managed to stand and hold up his bow but not draw it before the bear knocked his bow away and was on him. The father, aware of the commotion, ran to the scene and yelled at the bear, which then left his son and charged him. He got to full draw and let the arrow go at the charging bear when it was only ten feet away. The bear died inches before reaching him. Fortunately, they were able to evacuate the son by a helicopter, and he made a full recovery.

Accounts of similar black bear attacks on people are not hard to find. They have occurred in Minnesota, such as in 2002 when a University of Minnesota woodcock researcher was attacked near Mille Lacs. Most people I know are not terribly worried about black bears, but should they be? A study published by Stephen Herrero and his colleagues in the *Journal of Wildlife Management* provides some interesting details about fatal encounters between black bears and people.

When black bears have close encounters with people, they typically issue threats such as swatting the ground with one or both front paws, running toward the person but stopping short, or clacking their teeth; these behaviors may also be accompanied by "huffing, snorting, gurgling, and loud growling." These sorts of threats don't usually lead to physical attacks; instead the bear is posturing. If the bear has a way out, it will likely flee or let the person back away. Experts recommend not turning and running as this could cause the bear to attack, as it does with some dogs. Apparently this is some sort of "releaser stimulus" inherent in repertoire of predators.

At other times, fortunately much less common, bears make "predatory" attacks on people, which are sometimes fatal. In these cases, they are intent on killing a person for food, and they often drag the body away, bury, and guard it. In April 2003, a logging foreman in central Quebec was out surveying sites for timber harvest and didn't return that night. A search party the next day found his hard hat and boots amid a lot of bear tracks, and nearby his body in a den being guarded by a four-hundred-pound bear, which they shot. From tracks in the snow, they could see that the bear had paralleled the foreman, stalking, then moved ahead and attacked. Later tests found the bear healthy and disease-free.

Herrero and his colleagues began collecting data in 1967 on fatal bear attacks on people and covered the period from 1900 to 2009. They found that sixty-three people were killed in fifty-nine incidents by a wild black bear between 1900 and 2009. Of these deaths, forty-nine occurred in Canada and Alaska, and fourteen occurred in the lower forty-eight states. Although there were 3.5 times as many deaths in Canada and Alaska, there are only 1.75 times as many bears and many fewer people. Fifty-four of the fatalities occurred from 1960 to 2009. Some areas with high black bear populations had no fatal bear attacks. In case you're curious, the fatalities in the lower forty-eight states occurred in Washington (1), Wyoming (1), Utah (1), New Mexico (1), Colorado (3), Michigan (2), Tennessee (2), New York (1), Vermont (1), and Minnesota (1).

Fortunately, at least for interpreting any nightmares you may have, there is no evidence of a fatal attack in which more than one bear was involved—bears hunting in packs would indeed be terrifying. A surprise to me was that most fatal attacks by bears were by males, as I would have guessed females protecting cubs were the main culprits, as in the bow-hunting episode above. But males are larger and have larger home ranges, which potentially brings them into more contact with people (females tend to favor more secluded areas), and a male needs to eat more to maintain its strength and to protect territory. In Alaska and Canada, with lower food supplies and harsher winters, male bears might be more willing to consider a human as prey (they also have had less exposure to people), which could account for more fatal attacks. Herrero and colleagues noted that most fatalities occurred when only one or two people were present, and many of these were in August, when bears start fattening up for winter hibernation.

Many of the fatalities occurred when bears were attracted to people's food or garbage. People of various ages were killed and were participating in a variety of activities prior to being killed, including geological exploration, forestry activities, working on a drill rig, trapping, camping, working around a homesite, hiking, fishing, hunting, filming, jogging, horse riding, berry picking, and biking. And lastly, a candidate for a Darwin Award, one fatality occurred when the person was feeding the bear.

The authors suggested that people should learn to recognize predatory behavior in a bear that could signal its likelihood of attacking. In their words, "Potentially predatory approaches are typically silent, and may include stalking or other following, followed by a fast rush leading to contact." They stressed the value of carrying pepper spray and traveling in groups bigger than three when in bear country. They noted that a predatory bear may be deterred by shouting or hitting it with rocks, fists, or sticks.

Of course, if you don't recognize the signs of a bear's predatory behavior until the very last stages, just hope that you're not the slowest one in the group. In 2003, a 105-pound female hiker in New Jersey was stalked and attacked by a 400-pound bear, which she escaped from by smashing her elbow into his snout, stunning him long enough for her to escape! (Don't mess with those Jersey girls!) The increasing number of people and bears has and will continue to lead to greater frequency of encounters.

Obviously, your chances of being killed by a black bear are very low, and most encounters will not result in a bear attack. That there were only fourteen fatalities in the past 109 years in the lower forty-eight states was almost surprising. However, this number involves fatalities only, not attacks that resulted in serious injury. And lastly, note that the threat from brown bears both in western U.S. parks and in Canada and Alaska is far greater. A report I read noted forty-three deaths from brown bear attacks from 1960 to 2009 in a much smaller area. So, be vigilant and think ahead.

25

I WOULDN'T HAVE SEEN IT IF I HADN'T BELIEVED IT
A LOOK AT THE IVORY-BILLED WOODPECKER CONTROVERSY

I received a call from a person a few years back. He began by saying that what he was about to tell me would be hard to believe. Mind-blowing. He cautioned that I might consider him a kook, but in fact he prided himself on being a careful observer, and I should not doubt him. As curator of birds at the University of Minnesota's Bell Museum of Natural History, I was his first call.

He called to tell me that he had seen an Ivory-billed Woodpecker. In St. Paul.

The title of this piece is a phrase well known among birdwatchers and students of the outdoors. Looking through my binoculars, I too on many occasions have identified a rare bird, only to realize upon further review that it was something common. Granted, I have not talked myself into an Ivory-billed Woodpecker, but I've convinced myself of some real bloopers. How could I botch identifying a common bird?

Sometimes lights, shadows, and angles play havoc with your sense of color and perspective. Sometimes you see a bird out of context, in a place you don't expect it. Circumstances like these lead you to think you see at least one definitive marking of your rare species, and then your mind, fueled with enthusiasm of the moment, fills in what you know to be the other diagnostic features, leading you to your identification. Wrong identification, that is. It is quite amazing how the mind can trick us into seeing something so incorrectly, mostly because we want to believe it, and the error takes on a life of its own. This phenomenon is well known to attorneys, who must deal with the accuracy of "eye witness" testimony—sometimes witnesses simply saw what they believed.

Let's return to the Ivory-billed Woodpecker. What about the phrase "seeing is believing"? Quite a few folks believe they saw an Ivory-billed Woodpecker in the recent past. What is the history and current status of this bird? Extinct or not?

First, a bit of background. The ivory-bill needs large tracts of mature timber. In high demand also by the timber industry, these tracts were logged in rapid fashion. By the early 1900s, ornithologists and sportsmen and sportswomen alike were finding fewer and fewer of the big birds—so much so that even one sighting would touch off a flurry of activity by those wanting a glimpse. In the early 1930s, a state legislator from Louisiana, Mason Spencer, shot a male ivory-bill in a large tract of virgin timber known as the Singer Tract, along Louisiana's Tensas River. Word got out, and a few ornithologists visited the area and found a few birds. An artist, Don Eckelberry, went to the swamp in April 1944 looking for a bird spotted by someone else a few months before. He found a female at her roost hole and spent two weeks watching and drawing. Eckelberry's records constitute the last generally accepted sighting of an Ivory-billed Woodpecker in the United States.

After six decades without confirmed or documented sightings (many reports went unverified), a large woodpecker was identified by a lone kayaker as an ivory-bill on February 11, 2004, on a bayou in the Cache River National Wildlife Refuge, Monroe County, Arkansas. He led others to the site, and they claimed other sightings. This touched off a massive hunt led by the Cornell Lab of Ornithology. A poor-quality video was obtained, and putting this together with other observations, a team led by John Fitzpatrick from Cornell University published a paper in the journal *Science* announcing the rediscovery of the Ivory-billed Woodpecker. Was it true? Was the ivory-bill back from the dead? Had scores of hunters, fishermen, kayakers, bird-watchers, naturalists, and hikers missed this huge woodpecker?

The ensuing years saw a bitter battle between those who saw and those who doubted. One anonymous person provided new words to Ernest Thayer's 1888 immortal poem "Casey at the Bat" to ridicule the Cornell group. Unfortunately, an extensive search over four states and five years has failed to find any evidence that ivory-bills still roam the big woods remnants. Even the video has been scrutinized to the nth degree, and most are unconvinced that it is unambiguously an ivory-bill.

So to outsiders like me, the questions are, what did they see, and why have there been no subsequent sightings? Were the Arkansas sightings, not Eckelberry's, actually the last ones, and were the observers lucky enough to see the last individual of the species just before it perished? The prevailing opinion is that the observers saw an aberrant Pileated Woodpecker, a common bird throughout Minnesota and much of the United States, in particular, one with extra white on the wings, which would mimic the distinguishing characteristic of the ivory-bill. Pileated Woodpeckers occur commonly in the big woods, and extra white in the wings has been observed in some birds. One can imagine that given the fleeting glimpses, the bird could be mistaken for an ivory-bill, especially if one wanted very much to see one. Fitzpatrick and colleagues rejected this possibility.

Two scientific papers attempted to put some numbers on the probability that the bird still exists. One 2012 study by Nicholas J. Gotelli in the journal *Conservation Biology* concluded that the probability that the species exists is 0.0064 percent and that the last bird must have expired by 1980 at the latest! They used an interesting statistical exercise to investigate the probability of persistence. Say you want to know how many bird species exist in an area. You go to this place over and over and record every individual of every species every visit. After time you begin to see the same species, the common ones more regularly than the rarer ones. You reach a magic point when you have seen at least two individuals of every species. What this means is that the probability of there being a new species that you have not yet observed is extremely remote.

The Cornell team observed more than fifteen thousand birds, and each of the fifty-six species they reported were counted at least twice. The Gotelli team suggested that therefore the probability of observing a new species, that is, the Ivory-billed Woodpecker, was essentially zero. A second study, using different methods, arrived at the same conclusion: persistence is improbable.

At least some who reported seeing the bird stick to their guns. One experienced observer recounted that he got a ten-to-twelve-second look as the bird flew over the trees, and that the wing was white both on top and bottom, an ivory-bill hallmark. He commented that it flew like no other woodpecker.

Many attribute the lack of resightings to the bird having a large home range with plenty of places to hide. The believers continue to believe in the sightings. I was not there, but I have to number myself among the doubters. I do believe that those who reported the bird believe that they saw an Ivory-billed Woodpecker. I wish that were the same as actually seeing one. Still I hope I'm wrong and that somewhere, overlooked by armies of watchers, the species still persists. I fear, however, that believing is seeing, and our brief elation over the continued survival of the ivory-bill was a result of some optimistic mind's eyes.

26

RECENT DEVELOPMENTS IN THE CLIMATE CHANGE NEWS

Although I earned a Ph.D. in zoology, there is an enormous number of fields about which I know no more, and probably less, than the average person. Perhaps the most important thing I have learned is to recognize when I don't know enough to have an informed opinion. So if someone asks me what kind of rocket fuel NASA should use, I have to admit I don't have a clue, but I hope NASA consults experts and not people whose opinion is based on secondhand information.

I have been as interested as anyone else in the volatile field of climate change. I see at least two questions: (1) Has the earth's temperature increased in the past one hundred to two hundred years, and (2) if so, what caused it? Not being an expert in these fields, I have to rely on people who I think are experts based on their training and the research they have published in peer-reviewed journals.

Most studies suggest that the earth's average temperature has warmed. What does this mean, exactly? It does not mean that Minnesota will not have cold days or cold winters. It does not mean that the earth is warmer everywhere, all the time. Some areas might be colder. It's an average. And a change in the average temperature can result from several different scenarios. Low temperatures can stay the same, while high temperatures go up. Both lows and highs can get warmer. Or, as seems to be the case, the low temperatures can go up, while the high temperatures stay the same. In all cases the average goes up. To put it in local perspective, if winter temperatures are warmer and summer temperatures stay the same, the average Minnesota temperature goes up—over long periods of time, not every day, month, or year.

Numerous biological studies tell us that the breeding seasons of many animals in the northern United States and Europe have started earlier or their breeding ranges have shifted northwards. This is simply not debatable. A few of many examples will be enough. The midpoint in latitude of the American Goldfinch's range has shifted north by two hundred miles in the past four decades. Glacier lilies in Colorado bloom seventeen days earlier now than they did forty years ago. The range of the North American pika, a small mammal in the western United States with no tail and rounded ears, has shrunk by 50 percent in the past one hundred years. Pikas have very specific temperature requirements: they die in extreme heat and in winter if there is no snow under which to burrow. Pikas no longer occur in areas they once did owing to a changing climate. Similarly, those who have lived in Minnesota for long enough know that opossums are relative newcomers from the south.

Any one of these examples could be "explained away" without resorting to arguments about climate change. However, it's not rocket science to recognize a common theme in these very unrelated organisms. The earth's climate has gotten warmer. I can't think of a way to explain how numerous changes like the ones I just gave could all have happened for different, unrelated reasons, or by chance, which would result in an equal number breeding later in the season or shifting their ranges southward. So, even if experts disagree about what parts of the world are warmer or cooler and by how much, the message coming from biology seems clear. The earth is, on average, warming.

Still, I am taken by the large number of scientists actively studying climate change who have denied the trend. These folks largely concentrate on records from sources like weather stations, and they find evidence they think is contradictory; they do not focus on recent changes in the breeding dates and range shifts of plants and animals.

One such critic, Richard A. Muller, professor of physics at the University of California, Berkeley, a MacArthur Fellow and cofounder of the Berkeley Earth Surface Temperature project, has repeatedly denied that global temperatures have been increasing. In a recent statement, however, he reversed his opinion: "Last year, following an intensive research effort involving a dozen scientists, I concluded that global warming was real and that the prior estimates of the rate of warming were correct." In particular, he stated "that the average temperature of the earth's land has risen by 2 degrees Fahrenheit over the past 250 years, including an increase of 1 degrees over the most recent 50 years." Interested readers can view his group's scientific papers at BerkeleyEarth.org.

Muller's turnaround is especially significant given that his funding came from the Charles Koch Charitable Foundation, which has a long history of funding groups that deny climate change. So kudos to Muller for admitting that he believes the current evidence shows his previous stance was incorrect. Being able to change your mind and admit you were wrong is the hallmark of the scientific process.

Of course, some still deny global warming. For example, Marc Morano stated that "Muller will be remembered as a befuddled professor who has yet to figure out how to separate climate science from his media antics. His latest claims provide no new insight into the climate science debate." Morano is a former radio producer, not someone with his own personal scientific credibility.

But what about my second concern: What has driven the rise in average global temperature? First, let's take a step back in time. Fifteen thousand years ago a mile-thick glacier sat on top of parts of the upper Midwest. In a few short years, geologically speaking, the earth underwent a major warming that melted the glaciers and allowed life as we know it to return to Minnesota. So, we know that the earth warms and cools in major ways over time. The environment is probably never static, although during some periods change is less rapid. What can we attribute the recent rise in temperate to? Is it the last stages of glacial retreat?

The explanation that upsets so many is that we humans are responsible for global warming via our production of greenhouse gases, which insulate the earth and cause temperatures to rise. A look at charts of CO_2 emissions and temperature (see accompanying chart) shows us why many have suggested that the correlation is too strong to be attributable to chance. Muller came to agree with

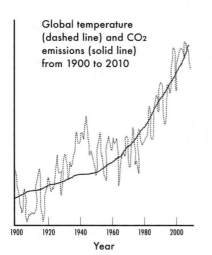

Global temperature (dashed line) and CO_2 emissions (solid line) from 1900 to 2010

Year

Plot of average global temperature and CO_2 emissions. Temperatures rise and fall over regular intervals, and unless we look at a sufficiently long period, we can find periodic declines in temperature. Many incorrectly use such periods as a basis to deny the relationship between CO_2 concentrations and rising temperature.

the majority of climate-change scientists on the cause of the global increase in temperature and remarked: "By far the best match was to the record of atmospheric carbon dioxide, measured from atmospheric samples and air trapped in polar ice."

Many of the implications of a continued rise in global temperatures have already been explored. For hunters and fishermen, the effects could be profound. The boundaries of wildlife management areas, parks, and wildlife refuges have been set in stone, whereas in fifty to one hundred years, the habitat that is today suitable for game species will have changed to one that does not support them. Their ranges will have probably shifted north of the current refuge boundaries, likely onto private lands without public access. Many species might be squeezed out altogether. Perhaps a good idea is to lease lands for public access and not buy them outright.

The final scientific story has yet to be told. Clearly, the earth's temperature has risen in the recent past. The cause is perhaps less well established. Some hold out hope that the rise in CO_2 and temperature will have a natural cause, that the correlation is just that, not a causal relationship. If I had to bet, I'd say that humans are the cause of the recent rise in global temperature, a rise that is probably more rapid than ones that have occurred naturally during the many ice ages, and is therefore more troublesome. As a scientist, I will keep an open mind, but I will continue to found my opinion on the work of people actively studying climate change.

NIGHT OF THE DEAD BIRDS, OR TOO MUCH HITCHCOCK?

My old friend Larry Conroy, who passed away at all too early an age, was fond of saying, "You know, the odd thing about rare events is that they sometimes happen." He would trot this out whenever some set of events occurred that people attributed to a mysterious cause. It is brilliant in its subtle sarcasm, but the point is too often lost on many of us.

I once received a call from another friend to ask my opinion about a recent spate of animal die-offs. The reports included three thousand blackbirds "falling from the sky" in Arkansas, Louisiana, Sweden, England, and New Zealand. Tons of dead fish washed up onshore in South America, crabs died off in England, and penguins and whales died mysteriously.

Mass die-offs are nothing new. Often in El Niño years, cold ocean currents result in the deaths of tens of thousands of seabirds. Nothing mysterious there. How often do you hear of whales washing up onshore? I do all the time. Standard stuff.

What about birds? We have known for decades that birds collide with man-made structures such as windows, communication towers, and vehicles. One of the earliest reports of such collisions described the deaths of 50,000 migrating songbirds that collided with a TV tower in Warner Robins, Georgia, on October 7–8, 1954. Yes, that's 50,000 in one night. Overall, it is thought that more than 100,000,000—yes, that's 100 million—birds die each year from collisions with buildings and other structures. Actually, that's the lower bound, the upper bound is 1 billion. So, biologically, these reports of sudden die-offs are neither new nor very impressive.

What in particular about blackbirds in Arkansas? Blackbirds flying about running into things and dying sounds Hitchcockian. But there's a reasonable explanation. Blackbirds roost in huge communal groups, numbering in the tens or hundreds of thousands. Many years ago, these birds and the common grackles they roosted with were targeted by various government agencies because they are agricultural pests (and might spread disease). They flew over the roosting birds at night and sprayed them with Tergitol, which causes the birds to lose the waterproofing in their feathers. When the spraying was followed by a rain, the birds died of exposure to cold—a wet bird dies quickly because it cannot maintain its high body temperature. In a well-publicized event in 1975, a roost of 3 million birds was sprayed, and an estimated 83 percent of the birds died. An outcry over this method of blackbird control resulted in its stoppage. But 83 percent times 3 million is a big number.

In the 1930s a late spring snow and sleet storm in southwestern Minnesota caused the deaths of tens of thousands of migrating Lapland Longspurs, small open-country birds that breed in the Arctic and pass through our state in spring and fall. Incidentally, because of the high reproductive potential of small birds like these, bird-watchers didn't notice any fewer migrating longspurs the next year. No one blamed anything other than a peculiar, relatively severe weather event.

OK, back to blackbirds in Arkansas. Blackbirds roost together at night, and they don't see well enough in the dark to fly, unlike many migratory birds. That's why blackbirds were vulnerable to spraying. If you disrupt the roost, say with fireworks on New Year's Eve, and cause a big flock to take off into the night sky, it is totally unsurprising that large numbers would kill themselves by running into wires, buildings, and maybe even each other. I find nothing suspicious or unusual about this. Is it unfortunate? Depends on your point of view—it might give the U.S. Fish and Wildlife Service a new plan for blackbird control. The population estimate for Red-winged Blackbirds in North America is 190 million. Three thousand dead blackbirds in Arkansas is 0.0016 percent of the total population.

I know that numerous theories about conspiracy and government involvement in these nearly simultaneous deaths have been concocted. Some will look for an insidious global virus, unleashed by a devious and evil group. Maybe aliens are to blame (although why they would have an interest in blackbirds, crabs, whales, etc. is unclear). But in the end, Larry was right again: the odd thing about cold winters is that die-offs sometimes happen.

28

EAGLE ATTACKS TODDLER!
THEN AGAIN, MAYBE NOT

I'm sure many people saw the video of a large raptor, suppos-
edly a Golden Eagle, flying through the sky, making a steep turn,
descending, and trying to carry off a toddler, to the dismay and
horror of the onlooking father. Several prominent television stations featured the relatively low-quality video, and it went viral on
the Internet, almost certainly causing panic attacks in people who
already fear the outdoors. This incident struck a chord with me
because as a kid I was blamed for losing track of a small dog during
a visit to a farm in southern Minnesota, whose loss was ultimately
blamed on a hawk attack.

Both are preposterous but show how far some folks have
strayed from a basic understanding of the natural world. The bird
in flight is not clearly identifiable. It could be an eagle, but the
terrifying scream that the bird utters in the slow-motion replay
is probably that of a Red-tailed Hawk, definitely not of an eagle
(their calls, which they rarely give, are described as high, weak,
and whistled). Here is the bottom line. It is basically impossible
for a raptor like an eagle to fly away with an object weighing more
than 50 percent of its body weight. An adult Golden Eagle female
(females are bigger than males) might weigh as much as twelve
pounds. So, doing some simple math, the kid would have weighed
six pounds at most, and as I recall, most kids weigh more than that
at birth. So a toddler? No way.

This also means that when you hear stories about eagles flying away with sheep, it's nonsense, unless it's just a part of a sheep.
The reason is that their wing loading is not designed to handle
that kind of payload. Reminds me of the time I was flying from
Newark to Minneapolis at Christmas time. The pilot announced
that we were being diverted off the runway so that we could rev

up the engines and burn off fuel so that we wouldn't be too heavy to take off. Same principle, and certainly got the attention of the passengers.

Raptors are, however, capable of killing prey too large to fly away with. One of the environmental services provided by our local Great Horned Owls is killing house cats that are let outside. Granted, the owls can't fly away with a large cat, but they can certainly kill one and eat their fill. The small lapdogs, same deal. Eagles can kill small deer and sheep, but again they can't fly away with them. They eat their fill and leave the carcass behind. And still other raptors will take advantage of a carcass—I often see Red-tailed Hawks picking scraps off of road-killed deer carcasses.

It was quickly ascertained that the eagle-toddler video was a hoax and was in fact a class project at a Canadian university. But I fear lots of people thought it was true. I'm sure they wanted the video to be a bit murky (like UFO sightings), because people would suspect a high-quality video of being a fake. It looked like someone was just "lucky" enough to be in the right spot at the right time to get the footage, with maybe a telephone camera. I give them pretty high marks, not just for the video but for preying on many people's lack of common sense about the environment. However, I have to go. I just saw a Sasquatch in my backyard.

IN THE WATER

29

RECREATIONAL FISHING ALTERS FISH EVOLUTION

I get depressed when I look at the price per pound for walleye at the grocery store, as it's often under $15 per pound. I calculate what it costs me to catch them, figuring in the boat, gear, gas, and lodging. Somewhere near $11.7 million per pound is my estimate. That might be a tad high, but it seems right.

Apparently, however, others are more successful at fishing than I, and the reason for this varying success may be rooted in fish personalities! The focus of a scientific paper by David Sutter and colleagues in the *Proceedings of the National Academy of Sciences* was the assessment of the impact of recreational fishing on largemouth bass populations and its effect on their potential for evolutionary change. Given the rate at which environments are changing, this would be useful information. They wanted to know whether differences among males in their willingness to attack a lure might have consequences for the population or species as a whole.

I had never considered the possibility that individual fish varied in how vigorously they would strike a lure. I always figured it had to do with water temperature, time of day, cloud cover, being hungry, or whether the lure looked fake, and that each fish was pretty much a carbon copy at the same sex and age. But apparently largemouth bass have personalities, at least in how aggressive they are. Some are more likely to strike a lure than others and are hence more vulnerable to recreational fishing. And this tendency is heritable! The fact that this behavior is heritable allowed the scientists to selectively breed two genetic lines of largemouth bass, High Vulnerability (HV) and Low Vulnerability (LV). Highly aggressive fish then are considered HV, owing to their willingness to strike a lure.

Sutter and colleagues focused on the breeding season, when male bass defend nest sites, at which females deposit eggs that the males fertilize. Male defense of the nest site consists of attacks on just about anything swimming by that might remotely be a threat, including scuba divers monitoring nests! Males also fan the eggs to ensure good water circulation and a sufficient supply of oxygen to the eggs, and males vary in how long they defend the nest during the season. Obviously, females want to choose males that increase the probability that her eggs will hatch and the fry will "fledge." In fact, a female deposits more eggs with a male she perceives as the most fit—that is, the most aggressive. But this aggressiveness includes a high willingness to attack a lure that gets too close to the nest.

Theory is one thing, testing it is another. The study involved six experimental ponds into which four HV and four LV males, a bunch of wild-caught females, and five hundred small sunfish (potential nest predators) were introduced. Males were monitored daily, and the time the male spent at the nest, the amount of time he spent fanning the eggs, and the number of days he defended it were recorded. That seems relatively straightforward, but how could you test whether they were more aggressive and hence vulnerable?

The investigators had the clever idea of casting three types of hookless fishing lures (surface popper, six-centimeter white twister jig, twelve-centimeter black plastic worm) right past the nest and watching how the males reacted. True to prediction, large HV males were twice as likely to attack the perceived egg predator (lures) than the wimpier guys. I asked Sutter if it wasn't just a matter of HV males spending more time on the nest and less time feeding, and being more likely to grab some fast food that came within easy reach. He replied that males usually don't feed when they are nest tending (meaning they need to come into the season in good shape). When I mentioned that none of the lures mimicked sunfish or some other nest predator, he said the males will attack just about anything, so that the type of lure usually didn't matter (although a surface lure might be ignored by a deep-nesting male). Large HV males also spent more time near the nest and actively fanned the eggs more than LV males. Lastly, large HV males guarded the eggs for a longer period of time over the season. Thus, it appears that it is to a female's advantage to seek out the big HV boys.

But these observations still do not nail down the answer to the question of whether HV males are more "fit" in an evolution-ary sense. Do females really get the most bang for their buck when they select the large HV males? The only way to tell would be to determine the paternity of the fry, which under natural conditions is pretty hard to do. That's where the beauty of the experimental ponds comes in. At the end of the experiment, the ponds were drained, and over eleven thousand fry were collected. A random sample of two hundred fish per pond was genotyped using stan-dard DNA technology and assigned to particular males. Because the data were not gathered by the LAPD, the result was clear: HV males fathered 62 percent of all the fry that were left at the end of the experiment (it is not clear how many the small sunfish might have eaten). The big HV males are indeed the top dogs, er fish.

Some males then are genetically predisposed to be highly vulnerable to fishing lures when nest guarding. And this is true not only during the nest-guarding season but year-round. Fish-ing removes these males, which are unfortunately those with the highest reproductive potential. But now we have the knowledge to make some clear recommendations. Many states start the bass season after males have finished nest guarding, as otherwise they could be easily located and targeted by fishermen. In some south-ern states, however, this is not the case. Bass tournaments where fish are taken to a weigh-in station should not be timed to occur during the nest-guarding phase, as significant nest predation can occur when the males are away from the nest (and they are likely returned in less than tip-top shape). For catch-release tournaments, one hopes they are returned to where they were caught, not miles from their nests.

These findings also have some interesting evolutionary implications. Ironically, heavy bass fishing pressure should favor LV males because they are less likely to be caught and removed from the population. However, in this case we could actually be selecting for less aggressive males, with the accompanying consequence of creating a pool of potential dads with subpar parenting skills. This then could change the bass population in an evolutionary sense.

If you could fish a lake that had never been fished before, you would experience some pretty good bass fishing, simply because the HV males would not have been already removed or reduced in numbers. The authors pointed out that these effects apply to any fish population subjected to recreational fishing. I now realize that all the places I fish for walleyes have been visited before by those good at catching the aggressive biters. I guess I just need to figure out how to catch the slackers.

30

DUCK HUNTING IN THE LOW COUNTRY; OR, HOW'S YOUR KOOIKERHONDJE?

On a recent trip to the coastal town of Gaast in the Netherlands, my host asked if I'd like to see a "duck decoy." I figured something was lost in translation, so to be polite I said, "Sure." Talk about a trip back in time.

This flat lowland area was originally treeless. Ducks frequented the area in winter and were fairly common but spread out. The ingenious means of concentrating and capturing lots of ducks in a short time originated in this area by at least the seventeenth century, and it's now called a duck "decoy," which comes from the Dutch word *kooi* for cage or trap. Here's how it worked.

First, a pond is surrounded by planted trees, to give passing ducks the illusion of a safe haven in an otherwise treeless landscape. The total size of the pond and the trees is only about 150 by 250 yards. On different sides of the pond, small channels about three yards across lead into the trees. This particular duck trap had six channels leading away from the center pond, like spokes. Covering each channel is netting raised up on poles, and along the channels is a series of fifteen-foot-long mat walls, or partitions, each about six feet high, which are arranged almost parallel along the channel but at sharp angles (see diagram). It is possible to walk behind the partitions toward the endpoint without being seen from the channel. The channels have at least one bend or curve leading to the endpoint; the point of each will soon become obvious.

The keepers raised ducks, called decoy ducks (often very noisy), so each pond had a supply of live "decoys." When the migrants visited this lowland area in fall and winter, they were naturally attracted to the tree-surrounded ponds and the live decoys. OK, so a bunch of wild ducks land in this pond. Did the

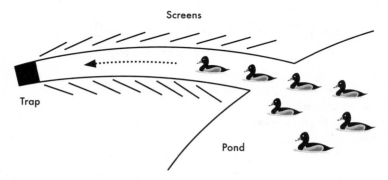

Schematic drawing of one of the channels leading away from the main pond.

locals open up and start shooting? Nope, these guys were doing this before there were decent guns. Plus, a few shots and it would be over.

How do you catch a bunch of wild ducks on the pond if you're not going to shoot them? The keeper lured the decoy ducks deeper and deeper into the channels by staying just ahead of them, out of sight, and throwing wheat or other grains on the water's surface. The wild ducks didn't seem to mind being invited to a free lunch, and what the heck, their new friends on the pond weren't afraid. I was thinking as this was explained, this is a good trick so far. You get a bunch of wild ducks heading down a covered channel—what next, get a net?

No, my host said, you use your "Kooikerhondje." OK, I'm lost again—not sure if I've got one, or if I do, whether it still works! The Kooikerhondje, I learned, is a small spaniel-like breed of dog whose job it was to stay mostly hidden behind the screens alongside the channel and help lure the ducks down the channel and under the netting. Now, I'm really curious, not to mention skeptical. How does a dog "lure" ducks? The Kooikerhondje has a really big, fluffy tail plume (like my English setter), and the ducks could occasionally see it as the dog went silently back and forth behind the narrowing openings between the partitions, following hand signals from the keeper. The ducks, believe it or not, were curious about the dog's tail plume and would keep following it down the channel away from the pond. When the ducks were far enough down the channel and around the bend and out of site of the main pond, the handler appeared.

That sounded a bit counterproductive to me because for my nickel, the ducks, especially the wild ones, would exit to the main pond at warp speed—game over. Turns out the decoy ducks were annoyed but just swam back out into the pond, as I predicted. The wild ones, however, didn't want any part of the Kooikerhondje-keeper action. Because they couldn't see the main pond, having gone around the bend, they flew or swam as fast as possible down the narrowing channel toward the light (and sometimes a mirror) at the end. For the wild ducks, the journey ended at the "catcher," a box where they were captured and killed. The decoy ducks were back in the main pond, having had a nice lunch, and were ready for the next unsuspecting group of wild ducks to appear.

The use of these duck decoys (as the locals call the whole operation) is now illegal for hunting. But some of the sites are still used for catching ducks to be banded for tracking and research, because the method is very effective. Some even still use the Kooikerhondje. Seeing it firsthand was a trip into the history of duck harvest I didn't know existed. There are lots of old duck decoys, and some, like the one I visited near Gaast, are maintained by dedicated volunteers as historical sites. If you get a chance to travel to this region, try to arrange a visit. You'll find it a fascinating experience. But remember, this area was enriched by lots of duck droppings, and I've never seen a bigger, taller, or denser patch of stinging nettles! Maybe it's the revenge of the ducks.

PREDATORS AND DUCKLINGS IN THE NORTH DAKOTA PRAIRIES

Minnesota duck hunters have witnessed a big decline in the number of ducks seen during recent hunting seasons. There are lots of possible reasons, such as migration routes shifting westward, too little food in stopover areas, or too many predators. Lots of waterfowl researchers are actively working on the issue, and I was interested in two studies published in the *Journal of Wildlife Management*, one by Matthew Pieron and Frank Rohwer, of Louisiana State University, and the second by Courtney Amundson and Todd Arnold, from the University of Minnesota. Their studies dealt in part with the role of predator removal on the survival of Mallard ducklings in North Dakota.

Now, as an aside, I confess that when I read discussions of whether duck numbers are down or up, I reflect on the past. We know that our skies were once filled with ducks, so many that even the market hunters thought (incorrectly) that their activities could not diminish duck numbers. If we think of the native prairies as nurseries of duckling production and realize that less than 1 percent of the native prairie remains, we can daydream about what seeing ninety-nine times more ducks would be like. Imagine being able to routinely pick out not just drake Mallards but drakes with triple curls. But that's irrelevant, as we can't go back in time, owing to the conversion of land to agriculture and our growing need for food. But researchers can help us identify how to raise as many ducks as possible given what we have left.

Pieron and Rohwer hired professional trappers to remove mammalian predators from township-sized study blocks (twenty-three thousand acres per block). Trappers were paid an average of $22,000, and they received a bonus based on nest success ranging from $600 for 30 percent nest success to $3,000 for nest success greater than 90 percent. (You've got to love the American way—an incentive-based trapping program!) Trappers removed 4,404 predators. Skunks (38 percent) and raccoons (35 percent) constituted the bulk of the predators; others included Franklin's ground squirrels, American mink, American badgers, red foxes, coyotes, and weasels. They were prevented from removing raptors by state and federal laws.

The main objective behind predator removal was to increase duck nesting success, and it was successful at that, increasing average nesting success by 1.4- to 1.9-fold, depending on the year. So removing these predators had a positive effect on how many eggs were hatched. But do more hatched ducklings translate into a larger fall flight? That's where Amundson and Arnold stepped in.

They studied eight plots in the Devils Lake Wetland Management District. Four of the plots were ones where predators were removed. There was little native grassland, except where land was enrolled in the Conservation Reserve Program, but the plots nonetheless had moderate to high densities of seasonal and semipermanent wetlands and relatively high densities of breeding Mallards and other dabblers.

Of the 168 broods they monitored, 109 (65 percent) experienced total brood loss—not a single duckling made it. Overall, they found one of the lowest duckling survival rates ever reported, from 0.7 percent to 34 percent across sites. Although their statistical analyses could not explain why the survival rates were so low, they speculated that it was due to the high numbers of mink and raptors, which were either not well controlled by trappers (mink) or not targeted by them (raptors).

Of those ducklings that did make it, being hatched early and being heavier increased their chances of making it to thirty days of age. Studies of birds typically find that being hatched earlier increases a bird's survival chances, as weather and food are likely more favorable and the bird can grow up before the change in seasons. Sometimes late-hatched young are just not ready to succeed under autumn conditions. Heavier young often do better because their survival doesn't depend as much on finding their own food during the first few days on their own, when they're learning what to eat and where to find it.

The important question is whether duckling survival rates were different on the plots with predators than on those without predators. The most astounding finding, even to Amundson and Arnold, was that "predator removal had no effect on duckling survival"; that is, there was no difference in duckling survival on plots with or without predator removal. Once again, they concluded that although the larger predators were removed or greatly reduced in numbers, a high number of minks and raptors remained on the predator-free plots and, they speculated, became major duckling predators.

Waterfowl managers often focus on duck nesting success, or bag limits, or hunting season lengths. Reading studies like the one by Amundson and Arnold makes me realize just how complicated it can be to figure out what factors are most important contributors to duck numbers. And doing fieldwork on this scope is daunting.

Predator control, amount of nesting habitat, size of wetlands for broods, and food availability likely all play roles, and they might interact in different ways in different places. For example, one might make the mistake of thinking that the answer is simple: more hens, more ducklings, more ducks in the fall. But Mallard pairs are territorial, and this territoriality can limit how many nests will be in a given area, meaning we need to increase upland nesting space. Also, too many ducklings can be counterproductive, as they will compete for limited food and at high densities might all perish, compared to a smaller number of ducklings that might all survive.

I wonder about the logistics of predator control on the nesting grounds of ducks in the midcontinent. At an average of $22,000 to trap township-sized blocks of twenty-three thousand acres, how feasible is this on a large scale? Likely not very.

Rohwer commented that having more coyotes helps duck populations because coyotes keep the fox numbers down, and foxes are major egg predators. I had assumed that coyotes also ate hens (if they could catch them), eggs, and young, and that is why everyone dislikes coyotes. But if more coyotes actually means more ducks, then perhaps duck hunters should boycott coyote hunting! Then again, in another essay, I reviewed research that suggested that fewer foxes mean more rodents, which help vector Lyme disease. Yes, nature has some complicated interconnections.

This topic has a few universal givens. Most people would like to see more ducks than we have at present. Various factors are keeping duck numbers below where they might be, and these could include the amount and quality of nesting habitat and cover, predator density, and food availability, to name but a few. The pressing, if not urgent, goal is to conduct large-scale research projects like the ones discussed here that can home in on the biggest culprits and then see if management can mitigate them.

32

LONG-TERM SEXUAL TENSIONS BETWEEN MALE AND FEMALE DUCKS

Once people learn I'm an ornithologist, I sometimes get motioned into a corner where I'm asked in a hushed tone, "Say, how do birds 'do it'?" Birds are at the opposite end of things like some worms, where "the act" can take seven hours. In birds, the act is not particularly dramatic, at least by Hollywood standards.

Most birds do not have a penis. Copulation is usually little more than a brief meeting of genital openings. Now, granted, some birds like Lapland Longspurs copulate 350 times per clutch (four or five eggs), so there can be a lot of action.

And courtship in birds can be a prolonged, expensive affair. Some males bring the female gifts of food, their version of our taking her out to dinner but, actually, more likely for illustrating the male's ability to find quality food, which will be crucial when bringing up baby. And some pairs go dancing, like Western Grebes, where the pair race across the water's surface in perfect choreographed unison, what David Attenborough called a pas de deux (a dance duet). Some males show off their value by growing showy plumages, often to the extreme.

Take the peacock. Not even a staunch adaptationist would suggest that the train of the male is for camouflage, tricking potential prey items, or aiding in aerodynamics. It is to impress the female, pure and simple. But how? There is evidence that the "eyespots" are an indicator to the female that the male is of high genetic quality, and in particular that he has resistance to disease or parasites. So to impress females more so than rivals, males evolve ever greater numbers of brighter eyespots, females choose them preferentially, and the process escalates in what evolutionary biologists call an "arms race." This process was termed *sexual selection* by Darwin (he got an impressive array of things right), and it occurs via the process now known as *female choice*.

But let's return to the basics of bird reproduction. I mentioned that few birds have a penis. In the past few years, our understanding of this part of the bird's anatomy has exploded. We now know that many ducks have a penis that is very long (up to the length of the male's body), twisted into a spiral, grows in the breeding season, and shrinks shortly thereafter (explaining why hunters don't notice them in the fall). The Peking Duck is a particularly good example.

These spiral-shaped penises function differently than those of most vertebrates. At rest, the penis is kept inverted (i.e., outside in). The penis literally explodes from the base under pressure from the lymph system (not blood supply like in mammals). This explosion, taking one third of a second, only occurs when the male and female genital openings are pressed together; that is, the erection (called an eversion by duck researchers) does not occur before mating. Sperm does not travel inside the penis but in a groove along the external surface.

Now, we can ask what the penis of ducks has to do with eyespots on the tail of a peacock. To answer this question, we first have to consider why some ducks would have a penis that is dramatically longer than "necessary" and spiral-shaped, when it would seem a simpler design would do. Like eyespots on the peacock's train, what's up with the penis of ducks (pardon the pun)? Does female choice play a role here?

Many have seen the aggressive behavior of Mallards in the spring. Two or more males are often seen in high-speed pursuits of a hen. One of the males is likely her mate, but the objective of the others is well understood, rape, or at least the bird version. If you see such a chase, it is obvious that the hen is trying to escape, as she actively tries to ditch her pursuers. Often, however, the males force the female to the ground or water and forcibly attempt to mate with her. On some occasions, the frenzy has resulted in the drowning of the hen Mallard. The terms used by duck researchers are *rape flight* and *forced copulation*.

Observations suggest that the female is not actively interested in mating with these other males. However, paternity studies have shown forced copulation to be an effective strategy on the part of males, as a proportion of Mallard hatchlings are fathered as a result of these rapes. Still, our theory is clear that females ought to be choosy about whom they mate with, and these flights are an obvious way to attempt to avoid unwanted pairings. But hens sometimes do not escape.

Back, again, to duck reproductive anatomy. Current thinking is that not just the male's anatomy is odd but the female's as well: the vaginal pathway is convoluted and sometimes has dead ends. Some researchers speculated that this could be a way for females to prevent unwanted inseminations; that is, instead of relying on a lengthy and potentially dangerous flight to avoid rape, perhaps her anatomy can help thwart unwanted matings. To know this, we have to know the intimate details.

What needed to be discovered was not going to come from observation. Some researchers decided to see if there were any consequences of the male's odd-shaped penis by experimenting! Patricia Brennan from the University of Massachusetts and her colleagues set up several devices designed to mimic a female duck's reproductive tract: straight tubes, spirals with clockwise and counterclockwise turns, and one with a 130 degree turn. Although they might sound "kinky," these were brilliant experiments!

They found that males were successful in everting their penises in the straight or clockwise spirals but not in the others. The replica of the natural anatomy could prevent the explosive extrusion of the penis. But, you ask, if they weren't successful with the device mimicking the natural female, why are there any ducks left? Here's where the analogy between the escape flights of hen Mallards and the reproductive tracts of ducks comes into play. Females of species in which forced copulations occur noticeably change the position of their body during mating depending on whether it is a wanted mating (i.e., with her chosen mate) or a forced attempt. If it is a forced attempt, they can prevent full eversion by assuming a particular position, and the sperm is deposited too low in the reproductive tract for fertilization. But when it's the correct partner, the hen assumes a position that allows full eversion and successful insemination. So, the researchers perhaps failed in that their apparatus didn't mimic a receptive hen but instead an unwilling participant.

The researchers speculated that the reproductive anatomies of males and females have undergone an "arms race" and coevolved. To prevent unwanted matings, female anatomy evolves a new twist, so to speak. To counter, male penis shape coevolves to be more successful, and so forth. What results is a reproductive system that is chock full of moves and countermoves and is very different from what one would speculate is "necessary"—very much analogous to the eyespots on the peacock's train. Females are in essence choosing particular aspects of male anatomy, and they evolve in response, just as Darwin predicted.

Not all ducks have this elaborate coevolved reproductive anatomy. Species that do not have forced copulations have no elaboration. But there are some new "twists" on the story. Recent evidence suggests that some male ducks actually grow longer penises when more males are around, making them the most likely one to father the ducklings. Once again, nature has been hiding some spectacular details from us. One of my ornithology students even suggested that my lecture on this topic should be titled "Pornithology."

As a postscript, in spring 2013 Brennan and her colleagues' research was attacked in the media by politicians who did not see the value in her research. Obviously it's easy to make fun of and challenge the need for federally funded studies of duck penises. However, the criticisms amounted to an ignorant mockery of science. Their work, funded by the National Science Foundation, passed a panel of experts, who evaluated the scientific potential of the work and awarded them a grant. Given that about 10 percent of grant applications are funded, this means the experts found their research to have a great deal of importance.

Not all research has to directly benefit humans. If you stifle the scientific process by placing such restrictions, we will all suffer. In hundreds of instances a discovery in one field has removed a roadblock to progress in another. Granted, not all research is deemed important by other experts, and one way this is determined is via the funding process. Evaluation of the value of research ought to be done by persons qualified to judge. In the case of Brennan and colleagues, their research was extremely insightful and valuable to the field of animal behavior, and their critics were badly misinformed.

VIGILANCE IN DUCKS
MORE THAN MEETS THE EYE(LID)

Several potential advantages accrue to individuals who live in flocks. An obvious advantage is being able to share guard duty. If you trust your flock mates, you can sleep or not spend a lot of time watching for predators as long as someone else is on duty, being vigilant. Then later you take your turn at watch, and over time, you are more efficient as an individual because of the sharing of guard duty among the flock members. Everyone must be "honest" for this to work. No slackers or cheaters allowed.

Although this strategy might seem simple, rules for flock living aren't necessarily straightforward. Researchers who study flocking behavior have made many insightful observations about how individuals change their behaviors in flocks of differing sizes. For example, imagine that you're in a small flock. Fewer eyes mean that you must spend more time watching for predators and less time feeding or sleeping. This would argue, seemingly, for being in a larger flock with more (potentially) cooperating birds. However, more birds in the flock usually means more fighting and perhaps less food per capita. Thus, there's a tradeoff between flock size and the advantage that accrues to individuals in terms of reduced guard duty. Bigger isn't necessarily always better when it comes to flock size.

Behavioral biologists are pretty sure this is true because of experiments with flocking birds and their predators. For example, if there were hardly ever any predators, a flock would be safe with just a couple of sentries. However, biologists can manipulate the apparent predator level by using either trained hawks or models. If flock-living birds perceive a greater-than-normal number of predators (compliments of researchers), flock size predictably increases, as more eyes are needed to maintain a certain level of vigilance. There is probably a certain distance at which a predator needs to be detected to provide enough time for the alarm to spread, the flock to scramble, take off, and confuse the predator. If the warning comes too late, the flock will have one less set of eyes, poor though they might have been.

Anyone who has watched a war or cowboy movie knows that the guy in charge tells the troops who has the first watch, second watch, and so on, and predictably one sentry falls asleep at a critical time (but that's Hollywood). How do birds decide who is on watch? Or for that matter, if you were a bird, how would you even know whether another individual was doing his or her job? I'd bet that most of us would pretend to be asleep and then sneak a peek to see if the appointed guard was watching out, loafing, or showing off.

A 2012 scientific paper in the British journal *Ibis* titled "Consistent Contrast between Eyelid and Iris Brightness Supports a Role for Vigilance Signalling in Ducks," by Matthieu Guillemain and colleagues, provided some interesting insights into the potential vigilance behavior of ducks. They observed forty-three duck species at a zoological park in Villars-les-Dombes in eastern France. These species included seventy-one plumage "morphs"— some of the species were sexually dimorphic, meaning that males and females had different plumages. The authors had the seemingly odd goal of determining whether the eye and the eyelid were contrasting in color.

At first glance, their goal might indeed seem like some arcane academic exercise. But their observations were exceedingly relevant to this question: if you're a duck in a flock, how do you know who is awake and hopefully scanning for predators and who is not? You need this information to determine your own vigilance level—is the threat level orange or red? We humans can tell, usually, if someone we can see is asleep, but can ducks?

Guillemain and colleagues observed that most ducks have a dark iris but a pale eyelid (and a third, nearly translucent eyelid called the nictating membrane). Therefore, if a duck is sleeping, the pale eyelid covers the eye, and it appears white to another duck. Also, the head color of many species (and morphs) is dark, further accentuating the pale eyelid, making it even more obvious whether a bird has its eye open. If the duck is awake and watching, the eye is dark. It's an easy and convenient signal: white eye means asleep, and dark eye means awake.

Of course what we often like most about nature are the exceptions. There are ducks with pale irises, and it was noted that most of these have dark eyelids. Only one species, the Rajah Shelduck (*Tadorna radjah*) has pale eyelids and a pale iris. However, this species has a white head, and in the images I checked online, the black pupil stands out rather clearly, suggesting it too can signal its vigilance level via the appearance of its eyes—pupil visible, duck awake.

The researchers pointed out that when a threat is noticed, more and more birds take a peek at neighbors and see the eyes-open status, and that this behavior spreads throughout the flock. The eyes may therefore also be a signal to ramp up the flock's alert status. Anyone who has approached a group of resting ducks knows that the number of ducks noticing the threat spreads throughout the flock like a wave from the point closest to you to the farthest side. Maybe a duck wakes up and sees a lot of open eyes, relative to its last check, and knows it's time to be on high alert.

But an alternative viewpoint needs to be considered. The notion that an individual duck will be at an advantage by taking turns at watching for predators has at least two critical assumptions. One is that the ducks are related, and hence this behavior may benefit close kin (birds that share some of their genes). A famous biologist, when asked if he'd risk his life for a drowning brother, said (tongue in cheek), "No, but I would for two brothers or eight cousins." That would be the same as "replacing" himself genetically (you share half of your genes with your siblings but only one-eighth with your cousins). We think that animals behave differently toward kin than unrelated individuals; that is, what appears to be altruistic behavior is actually a form of self-help if directed at kin.

Alternatively, if the birds are not related, the only way this behavior could succeed without cheaters profiting is if birds reciprocate honestly. I contacted Guillemain, and he expressed this view: "Conversely, I believe each individual will try to spend as little time being vigilant as it can and use the vigilance of others as much as possible." In other words, he thinks that the ducks are more likely parasitizing rather than sharing each other's behavior. In the case of ducks, showing clearly that your eyes are closed may be a way of saying, "Hey, I am not a reliable guard, so spend more time on vigilance yourself," hence pushing your neighbors to be more vigilant, so that you can spend more time sleeping.

Behavioral biologists also know that in nature, cheating can be profitable. In an innocuous human analogy, haven't you gone to dinner with a group in which one person orders something way more expensive than everyone else and then wants to split the bill equally?

Thus, what the researchers showed is the potential for waterfowl to be able to signal their vigilance status by means of the contrasting iris and eyelid coloration. But we have much to learn about the function of the iris and eyelid colors. As we know from hundreds of studies, what we think is a likely a function of some structure is often wrong. So at this point it's good theory, and it opens the door for others to watch variously sized flocks to see if the number of birds on guard duty matches theoretical predictions.

THE THREE-MINUTE OUTDOORSMAN

As an alternative, for example, one might suggest that the iris and eyelid coloration functions in the way in which ducks choose mates, sort of a Morse code for dating. Two winks means yes, one wink, maybe. However, in this case, the researchers noted that ducks have their eyes open during courtship. That right there might be a decent take-home lesson for humans.

34

WHAT LITTLE WE KNEW ABOUT THE LABRADOR DUCK JUST GOT LITTLER

We are all aware that the days of ducks filling the skies are long past. Still, seemingly good numbers are around, and none have recently gone extinct. This is not the case for the Labrador Duck, which was extinct by the late 1800s. Our knowledge of this bird is pretty fragmentary. For example, Audubon may have seen a nest of the species (in the nonbreeding season), but whether the nest he described is actually from this species is debated. So, no one is actually sure where the species' breeding grounds were ("Labrador" was kind of a guess). We know from museum specimens (fifty-five worldwide) that it wintered along the Atlantic Coast, from Nova Scotia south to Chesapeake Bay. A substantial number were collected near Long Island, New York. Some described the duck as common, others said it was rare, while still others described it as common at one time and then rare as market hunting reached its heyday. Some found the duck to provide an "excellent supper," whereas others said the birds often hung in the New York City markets until they rotted. In any case, delectable or otherwise, they are extinct today. What caused their extinction is unclear. They were hunted for food on the wintering grounds, and indigenous peoples may also have eaten them where they bred. In the 1700s and 1800s, feather and egg hunters may have made an impact. The duck also had a very specialized bill, which

132

probably meant its food source was highly specialized and possibly made the species vulnerable.

So, our knowledge of the Labrador Duck involves old, often conflicting reports and museum specimens. Until recently, these specimens included both museum skins and eggs. However, eggs of many species are not that easy to tell apart from the empty preserved shells in collections. So, the eggs thought to be from Labrador Ducks were perhaps misidentified. But given that the bird is extinct, how would we know?

In a scientific study published in the ornithological journal *The Auk*, Glen Chilton, of St. Mary's University College in Calgary, Alberta, and Michael Sorenson, of Boston University, tested Labrador Duck eggs to see if they were authentic. Chilton and Sorenson searched museum collections around the world and found nine eggs that were identified as those of the extinct Labrador Duck. Chilton extracted some of the dried egg membrane from the inside of each of the nine eggs (each at least 150 years old), which were in Germany (six), England (two), and Scotland (one). He sent them to Sorenson's laboratory at Boston University. Sorenson then used some careful laboratory techniques to extract and sequence DNA from the membranes. It is well established that each species has a unique DNA fingerprint or profile, and Sorenson compared the sequences from the eggs shells to DNA sequences obtained from some of the old museum specimens (called study skins) of Labrador Ducks as well as other waterfowl species.

The results were not good if you wanted to know what Labrador Duck eggs actually looked like. The DNA tests showed that the six eggs from Germany (Dresden) were produced by one or more Red-breasted Mergansers. The two eggs from England were produced by a Common Eider and Mallard, respectively. The Scottish egg was also from a Mallard. In every case, the range of the species that actually produced the egg overlapped with potential Labrador Duck breeding sites. Unfortunately, none of the eggs were from Labrador Ducks.

Some of the confusion about the identification associated

with the old eggs comes from the common names people have used for birds. For example, the name *Labrador Duck* was sometimes used as a name for a domestic, Mallard-like duck now known as the Black East Indies breed. So, an old common name associated with some old eggs gave the mistaken impression that one of the English eggs was from "the" Labrador Duck.

In any event, thanks, I guess, to modern technology, we now actually know less about Labrador Ducks than we thought we did. I guess it's better to know this than to think we had a few more glimpses into the biology of this long-extinct species. Certainly the museums now can correctly identify the eggs, although they might be displeased that they were possibly duped long ago into buying eggs that were either honestly misidentified or deliberately falsified. The museum in Germany (Dresden) purchased at high cost their six "Labrador Duck" eggs in July 1901—now it's time to update their catalog!

MUMBLING ALONG
LESSONS FROM THE PAST ABOUT STOPPING THE SPREAD OF EXOTIC SPECIES

The spread of aquatic invasive species has become epidemic. Departments of natural resources struggle to stem these invasions, although their efforts sometimes seem too little too late. Many procedures have been introduced to reduce spreading from lake to lake. Removing aquatic vegetation that hangs from boat trailers is one. Others are to drain live wells and bait buckets and to remove a boat's plug when leaving a lake or river. If these actions are not taken, invasive plants and animals harbored in these small amounts of water can be easily exchanged between bodies of water. I have pulled several pieces of debris from the bottom of the St. Croix River with zebra mussels attached.

These minor inconveniences are actually pretty annoying; just ask the many who have forgotten to put the plugs back in their boats before the next launch, after decades of not needing to. But these inconveniences pale when we see underwater pictures of large, spreading beds of exotic zebra mussels on the bottom of the premier walleye fishery Mille Lacs in central Minnesota. Shortly after their introduction, they spread, figuratively speaking, like wildfire. And they have an impact, as I found in the summer of 2013 when I visited Mille Lacs and observed how much clearer the water had become, a known effect of zebra mussels. Although clearer water might seem a good thing, it's not if that's not the lake's normal water clarity.

But what's the big deal? Having a new species on the landscape is not a novel occurrence; it happens in nature all the time. Over scales of thousands of years, entire communities of plants and animals have moved great distances, such as the wholesale recolonization of areas like Minnesota and Wisconsin following retreat of the last glacier. We also see ranges shift over the short term. Many birds, like the little Blue-gray Gnatcatcher, have markedly extended their ranges northwards in the past twenty-five years. And who in Minnesota needed a recipe for road-killed opossum thirty years ago?

The problem of introduced species is often greatest when the exotics come from faraway places and have some advantage in their new environment: they may have adaptations that make them more successful than the local, native species, or they may be preadapted to a man-made niche that only recently became available. In addition, introduced species typically arrive without their typical predators, and competitors are lacking. Carp are a prime example, and we have known for years that they can harm aquatic environments.

Only recently did I see some figures that put this concern in perspective. And the information isn't new; it comes from an article published by Alvin Cahn in 1929 in the journal *Ecology*. He told a fascinating story about a small lake in southern Wisconsin and its tale of woes from introduced carp. It reinforces that we should be concerned about introduction of exotic species, and it strengthens the conviction that we should proceed as though it isn't too late.

Cahn's story begins with the carp of Pike Lake, a widening of the Neosho River in southern Wisconsin. Pike Lake was seminatural in that it was a natural widening of the river, but it was later made more lake-like by a dam. According to Cahn, largemouth bass and northern pike were easy to catch there. Farther downstream, the Neosho River goes through the town that gave it its name, and eventually enters the Rock River. (Incidentally, for those familiar with this area of Wisconsin, some of the names of places and routes of rivers have changed.)

Around 1880 the citizens of Neosho constructed another dam to support a local mill for grinding grain. This lake, called the Neosho Mill Pond, also became full of fish from upstream and supported a wealth of native aquatic vegetation. It was a popular destination for local fishermen, and many of the same species from Pike Lake occurred there.

But, and this will be all too familiar, Cahn noted that in Pike Lake, fishermen used carp minnows for bait and at the end of the day dumped the living ones into the lake. Now, when I was a kid, I figured that dumped minnows wouldn't survive and we'd just be giving an offering to those fish that escaped our considerable fishing skills! But this was not the case, and many carp minnows survived, got bigger, and began to reproduce. Remember, carp are exotic and behave in ways that allow them to escape our otherwise efficient native predators. Carp became fairly common. This was not good.

After time, the dam supporting Pike Lake failed, and everything ended up downstream in the Neosho Mill Pond to the south, including large numbers of carp. The pond was actually better habitat for carp, as it had a muddy bottom, and they proliferated. Fishing for the usual species started to fail and deteriorated to the point where the only fishing was for carp by setlines. No one bothered to fish for anything else.

In September 1924 the dam supporting the millpond started to fail, and it was decided to drain it while constructing a new dam. This provided Cahn the opportunity to evaluate the effects of the introduction of carp on the once-great local fishery. To me this is key, as not being an aquatic animal, I confess a high degree of ignorance as to what really lives below my boat's keel.

A net was placed across the river, and the pond was drained. What remained were large mudflats (which were originally dry land before the dam) and the water-filled channel that was the original Neosho River. As the waters receded, Cahn noticed two striking features. First, there was no aquatic vegetation. None. Second, the muddy bottom of the pond was pitted. In Cahn's words, "everywhere were moon-shaped or semi-round depressions about a quarter of an inch deep," and he mentioned that many of the depressions overlapped.

The depressions are the work of carp, created as they press their mouths on the muddy bottom and suck up vegetation, a behavior called "mumbling." The carp had rooted out every piece of aquatic vegetation in the pond. The water was no longer clear—it was muddy from the mumbling activities of carp. What about the fish he observed?

"When the first seine came in I had visions of seeing some beautiful fish, but the first draw disillusioned me: there was nothing but carp," wrote Cahn. The full results of his netting are staggering. Ninety-eight percent of all fish in the pond were carp, a whopping 37,750 pounds when shipped to market. The few northern pike and walleyes were skinny and, not surprisingly, "full of tapeworms."

The fish numbers from Neosho Mill Pond are even more depressing when you see Cahn's results from nearby Little Silver Lake, a lake with similar characteristics (third column in the table on next page). Notice that the number of sunfish (both species) in the pond was 0.1 percent of the sunfish it should have had. Can there be any clearer evidence for preventing the spread of carp or any other invasive exotic species?

TOTAL NUMBERS OF FISH FOUND IN PONDS WITH AND WITHOUT CARP

Species	Neosho Mill Pond (with carp)	Little Silver Lake, Waukesha Co., Wisc. (no carp)
Carp	5891	0
Buffalo fish	1	0
Smallmouth redhorse	66	0
White crappie	17	0
Redhorse	14	10
Walleye	4	20
Northern pike	3	380
Bowfin	7	340
(Blue) sunfish	2	1220
Rock bass	1	940
Largemouth bass	0	1120
Pumpkinseed sunfish	0	610
Black crappie	0	730
Yellow perch	0	680
Longnose gar	0	30

The devastation by carp of the native fish of Neosho Mill Pond was complete. It isn't clear to me how the carp survived given that they had eaten everything, but they averaged 6.6 pounds. Now, over normal evolutionary time scales, some predator or parasite would exploit the overabundant carp, and some new equilibrium would be reached. Maybe with long-term restoration, something resembling the pre-carp fishery could have been established. But that wouldn't do much for the immediate fishing prospects.

Today, we quickly forget these sorts of lessons, but they remind us that our aquatic and terrestrial habitats are fragile. They have evolved a delicate balance over millennia, and it can be relatively easily upset. This knowledge should serve to reinforce our efforts to prevent exotics from being introduced or spread. We should take to heart the lesson of Neosho Mill Pond. We should fully support efforts to stop invasions of new carp, such as bighead and silver, and to try as best we can to eradicate the common carp that are already established. In the end, a few minor inconveniences like draining our live wells is but a small price.

36

WHAT YOU DON'T SEE UNDER YOUR BOAT

One my favorite sayings from philosopher Delos McKown is "the invisible and the non-existent often look very much alike." Of course, not everything invisible to us is nonexistent, but some of the things we cannot see, at least with the naked eye, are pretty insidious. Think Ebola virus.

We are not unaware of microscopic health hazards lurking in water, as everyone who has camped and boiled water knows. These health hazards, like algal blooms and bacteria, are now fairly well-known. A less well-known hazard is air pollutants, such as mercury, that can be washed into our rivers and lakes. The burning of coal is the source of most of the mercury pollution of the air. Mercury used in products, including fluorescent lamps and fillings in our teeth, can also be released to the air and find its way into our waterways. Hence, we often have fish advisories owing to the fact that fish "bioaccumulate" mercury in their tissues, making them unhealthy to eat, at least in large amounts. For example, only one meal per month of white bass from the St. Croix River above Stillwater is recommended by the Minnesota DNR. In recent years we have become more aware that what we dispose of via wastewater is not "gone" once we flush our toilets, finish watering our lawns, or drain the kitchen sink. For example, the drugs we take are not completely "used up" by our bodies and actually end up in the aquatic ecosystem. More specifically, when we take a prescription drug, or even something like ibuprofen, our bodies do not completely break it down, and we end up excreting some of it. Wastewater treatment plants do not remove these substances or have tests for their presence, and therefore they are still present in the "treated" water that is released back into our environment.

An article in the journal *Science* in 2012 by Tomas Brodin and colleagues had the eye-catching title "Dilute Concentrations of a Psychiatric Drug Alter Behavior of Fish from Natural Populations." Here they established that personality-altering drugs designed for people can alter the behavior of fish, European perch specifically. How's that, you ask? So did I.

This study looked at the effects of oxazepam, one of a group of antianxiety drugs called benzodiazepines. The drugs act on a common type of nervous-system receptor found in a wide array of vertebrate animals, including mammals like us and fish. The team (from Umeå University, Sweden) found that oxazepam was easily detectable in the Fyris River. European perch from that river had tissue concentrations of oxazepam six times higher than in the water, showing that the fish "bioaccumulate" the drug. Not good news for perch.

The researchers wanted to know if the drug had any behavioral effects on the perch. They put one group of young perch in water with similar levels of oxazepam as was found in their natural environment, and put a second group in clean water. The oxazepam group showed drastically altered behavior! The perch were less afraid to venture into new areas, generally swam away from other perch, and were more active and quicker to eat zooplankton.

If you're a fish, leaving your school is usually a bad idea, as by yourself you're more likely to be picked off by a predator. Perch do eat zooplankton, but if they eat too much, then the zooplankton don't eat as much algae, and algae populations soar, disrupting ecosystem balance. So, maybe it isn't a huge deal for Minnesota perch yet, but even a small amount of oxazepam in the water could negatively alter fish behavior. I guess I think that our fish have enough problems already without further altering their behaviors.

Oxazepam is not the only unnatural substance in our waters known to alter fish behavior. Some of the components of antibacterial soap (triclosan and triclocarban), for example, have been found in surface waters and were found to affect the behavior of fathead minnows. In particular, wild male fatheads exposed to the chemicals were less likely to defend their nests against artificial minnows than the males in chemical-free water were. I have washed my hands with this type of soap, and I knew that overuse of it allows bacteria on our hands to evolve resistance, but I wasn't aware of these other negative effects "downstream."

A study by Dalma Martinović-Weigelt from the University of St. Thomas in St. Paul, Minnesota, found that low concentrations of ibuprofen, which reduces inflammation in people (and is one of my preferred drugs for pain) influenced the behavior of zebrafish. So, instead of helping a fish get over a hangover, it makes their sexual behavior stray from normal!

But, what the heck, no one eats zebrafish. And how widespread can this problem be? Unfortunately, possibly a lot. The Minnesota Pollution Control Agency recently surveyed waters upstream and downstream from sewage treatment plants (point sources for where drugs and other contaminants could be released into our waters). They found antibiotics, a drug used to treat attention deficit hyperactivity disorder, antidepressants, chemicals from detergents, and contraceptive hormones (known to de-masculinize some fish, and in others to produce individuals with both male and female structures!). These substances of course do not occur naturally in our waters.

Most of these substances occurred in minute amounts, which would be good news except for the fact that for many of these drugs, a little goes a long ways in terms of eliciting a response. Clearly, it is good that this potential problem is on our enviro-radar and that many scientific studies are examining the potential effects in wildlife and humans. Vigilance then, and forewarned is forearmed.

NEVER BE A BABY BIRD

I learned ornithology from Dwain Warner, a longtime curator of birds at the University of Minnesota's Bell Museum of Natural History. Dwain, or DW as we called him, had a number of sayings, but one of his favorites was "never be a baby bird." He was referring to the fact that the fate of most baby birds is to perish at a young age, maybe even before hatching from its egg. For example, about 90 percent of Black-capped Chickadees never make it to their first breeding season. Hence, DW's admonition to avoid being a baby bird, and since I now have his position at the university, I feel compelled to keep up the tradition.

Baby birds are extremely vulnerable, especially those that are altricial, meaning the young are hatched essentially featherless, with eyes closed, and basically 100 percent dependent on the adults. The babies are essentially living guts with a head and legs. When a predator attacks, it typically wipes out the entire nest. I taught a field ornithology class at the Itasca Biological Station, and we once watched a red squirrel go into a nest hole in a dead tree and bring out all of the baby Black-capped Chickadees one by one, eating them like they were ears of corn, partially grown feathers raining down. All the while the adults were scolding the squirrel, who wasn't bothered in the least.

A different way to be a baby bird is to be precocial. Species like ducks and turkeys have babies that are basically ready to go right out of the box (egg). They can see, hear, follow mom (imprint), run and hide, and mostly feed themselves after a short period. If you've followed a group of Canada Goose goslings in the spring, you may have noticed that, unlike in groups of altricial birds, the number in the little flotilla decreases slowly over time—sometimes one less gosling a day. When the young are able to scatter when the parent sounds the alarm, predators typically don't get the whole bunch at once but have to pick them off one by one. But if you put all of your eggs in one basket (altricial birds), a predator can easily eat them all at one setting.

But in the end DW was still right. We are not neck-deep in birds, and the reason is that baby birds, whether altricial or precocial, find it really tough to make it to adulthood. Predation prevents populations from growing out of control. Now, I can name a number of predators that would take baby goslings or ducklings, but Lisa Dessborn and colleagues asked a slightly different question. They wondered not which predators would take baby Mallards but how Mallard ducklings react when confronted with potential predators. Do they have some built-in ideas of what to avoid because it might want to eat them? Part of the motivation for their study was that researchers would have to spend a ton of time observing ducklings in the wild to see many predatory attacks on them, so why not ask the Mallard ducklings who they're afraid of? They published their results in the journal *Behaviour*.

The authors bought Mallard ducklings and introduced them to a series of fenced enclosures surrounding small ponds in Sweden. They created broods by keeping birds together (a bit of paint on their backs helped the researchers know who belonged with whom). They exposed the ducklings to various threats and devised ways of scoring their responses. First, they played calls from crows and gulls, known duckling predators, and for comparison, they played calls of native species, like finches, that are not threats. Two dead pike were purchased from a local fisherman (says something about the researchers' fishing skills?). The broods were removed, and the (frozen) pike was placed in the center of the pond. They also simulated a pike attack by attaching the dead pike to a fishing line and rapidly moving it toward the brood. To simulate aerial attacks, they "flew" a stuffed goshawk over the ponds along a wire. Unfortunately, they didn't have "controls" in either of the last two experiments. The researchers ought to have flown something like an oriole or a milk bottle over the pond and suspended a nonthreatening aquatic creature or a bottle of water in the center of the pond.

Lastly, because the introduced American mink are hugely successful duckling predators, a captured mink (that had killed a brood and parts of two others) was put in a cage next to the pond, in view of the ducklings.

I was really intrigued by this study. What behaviors for predator avoidance would the ducklings have hardwired? When they heard the recorded bird calls, they swam to the middle of the pond and craned their necks, with "apparent vigilance." However, their response was much stronger to the predatory birds, showing an innate response to a potential aerial attack. When the birds saw the goshawk coming toward them, most but not all of the broods dove and scattered and surfaced some distance away. I was surprised the response was not stronger. They also noticed that some ducklings seemed to be good followers, reacting to other's behaviors.

Oddly, the birds were not terribly concerned about the mink in the cage, although they did tend to stay on the other side of the pond. The broods were also apparently nonplussed by the suspended pike. This was true even after the broods had been "attacked" by the pike. Maybe they "sensed" that it was frozen, and I don't take much from this observation.

The simulated pike attack, however, was a different story: "As the pike dummy broke the water surface, ducklings in all broods scattered about by running on the water surface and onto shallow waters, and two ducklings in one brood even ran up on land." Now that meshes with a similar attack I saw on a brood in northern Minnesota—chaos in the water.

So, the results were not as straightforward as I thought they would be. I was most "bothered" by the lack of response to a motionless pike, apparently sitting in ambush. So were the authors, as they pointed out that pike are widespread, they eat ducks and ducklings, and ducks do not avoid lakes that contain pike. The authors suggested that ducklings are unable to identify pike and other large fish or even similar-shaped objects when under the surface. I personally would have seen if they found a snapping turtle decoy similarly unthreatening.

There is an obvious trade-off here, in that ducklings cannot avoid water; in fact, we think it helps them avoid land-based predators. Furthermore, they feed on the water's surface, and they can't just not eat to avoid being eaten themselves. So maybe part of the strategy is waiting until a clear threat, like a large fish, draws near before scattering. And we should not overlook the fact that "scattering" is likely a strategy to confuse a predator, potentially causing a moment's hesitation, resulting in everyone's escape. I subsequently saw a home video on YouTube that showed this exact result when a northern pike in Alaska attacked a Mallard brood from underwater.

The authors also commented on the obviously unexpected result of very little response to the mink in the cage. This study was done in Sweden, and they suggested that since ducks in Europe have not evolved with a predator of this kind, perhaps local ducklings lack the appropriate response. That I doubt, but I don't have a good explanation either. Someone ought to import some eggs from North America and do the same experiment.

In the end, DW's pronouncement still rings true. Even with the ducklings in captivity, a mink broke in and ate a bunch of them. So, altricial, precocial, or in between, never be a baby bird.

38

OH, NO! DUCK HUNTING VIDEOS MIGHT NOT BE REALISTIC!

I spend a lot of time watching duck, goose, and archery deer-hunting videos because they cut the boredom of my treadmill sessions. My unofficial opinion is that in deer-hunting videos, there's a 95 percent chance of a shot and kill. If I took that literally, I'd have to call myself a terrible hunter, because my shot chances per hunt are more like the opposite. I used to be amazed that some hunts were shown on these videos because they had what I thought was very poor shot placement. However, I finally learned that they don't show the hunt if they don't have footage with the hunter and dead deer. Some video producers who are starved for impact shots will stoop to showing pictures of a faraway buck or a miss and claiming "that's hunting," but it was obviously not their intent. The team didn't connect and had to have some footage to fill a show.

Honestly, I think everyone knows that hunting videos stray from realistic to ultrasuccessful. That's why we watch them. But just how unrealistic are they? A 2013 paper in the journal *Human Dimensions of Wildlife* by Mark Alessi and colleagues, titled "Content Analysis of Three Waterfowl Hunting DVDs," gave some real numbers.

First, they reviewed the role that visual media play in our lives. Kids two to eleven years old watch 117 hours of TV per month, and adults eighteen to thirty-four years old, 131 hours. This provides plenty of opportunities for young minds to be molded by the content of what they watch. Products and services are pitched, often with deft subtlety. Many young kids are no match for the creative minds of advertisers who want to convince them they're not good enough in some way or that a product gives their friends an advantage.

But are hunters influenced by what they see on hunting videos? There is evidence that deer hunters that watch hunting videos have a preference for harvesting trophy bucks. What about waterfowl hunters? More kids live in urban areas, and thus they are less likely to experience a real hunt, and more likely to think that what they see on a hunting video is what it's always like. Constant action, ducks falling everywhere, dogs exhausted from countless retrieves. Say it ain't so!

But that is exactly what the videos showed. The authors ran the videos through a word analyzer and found that the most common words were *kill* and *ducks*, but not always together. Only 3.8 percent of each movie involved actual shooting at ducks (that's pretty realistic; I spend more time drinking coffee). The actual length of each shooting scene was less than two seconds. Now here is where the steel meets the sky.

On average, 52 percent of the scenes involved one to three shots, 36 percent involved four to ten shots, and 10 percent showed more than ten shots. We all know that shots don't equal ducks. Only 3 percent of the scenes showed no ducks killed, 85 percent of the scenes showed one to three ducks downed, and 13 percent showed between four and ten ducks harvested. Overall, ducks were harvested in 94 percent of the scenes, but in general, one to three shots were fired and one to three birds harvested per scene. Nope, it ain't so.

The authors noted that in 2011–12, Illinois duck hunters harvested an average of 0.7 ducks per day, 53 percent of hunters harvested less than 6 ducks, and 15 percent harvested no ducks the entire two-month season. Frankly, I've had a couple of adventures where two of us in less than an hour shot eight-duck limits in Canada, but these are rare times (I would not care to say how many shells were used). Such very successful times are far from the norm. The authors made the obvious conclusion that "the high probability of success (94 percent) conveyed in the hunting videos is not a realistic representation of success in waterfowl hunting." They further added that they didn't know if these videos would lead inexperienced waterfowl hunters to think that a success rate of over 90 percent per "scene" is realistic.

But other than unmet expectations, could there be a down side to these videos? The authors say that some hunters blame these videos for higher crippling rates. If they think they should be killing more ducks, maybe newer hunters are driven to shoot at birds out of range.

An interesting but almost side note was on shooting banded ducks and geese. It was estimated that there were 9.2 million Mallards in the 2011 season, of which 15,605 birds were banded that year. However, the 9.2 million figure is for the breeding population and does not include young birds. The fall flight was probably something closer to 20 million (thanks to Dr. Todd Arnold for that figure). In addition, because not all banded birds die in their first year, there are more banded birds in the population than were banded that year. In 2010, 4.16 million Mallards were harvested, of which 15,825 were banded (thanks again, Todd Arnold). Also, Arnold mentioned that some feel that the reported harvest is inflated and might be closer to 3 million. Still, the probability of harvesting a banded bird is very low, yet each of the three videos showed the relatively improbable harvest of a banded bird (one video showed a hunt in which thirteen banded geese were taken).

Newer hunters might conclude that hunters with lanyards full of bands are true duck-killing masters and that banded birds are trophies. The authors suggest that such videos might lead to a preference for shooting banded birds, which could frustrate duck biologists. Although banding is useless without recoveries, many of the statistical analyses assume that mortality is random with respect to banding, and preferentially shooting banded birds could bias such studies. For example, if banded birds were targeted specifically, they would have lower survival rates (and shorter lifetimes) simply because they were banded. Nonetheless, bands recoveries should always be reported.

In this era of instant gratification, I think that these videos might in fact give new hunters unrealistic hopes. I enjoy watching them, but I do think they plant in the back of my mind the hope-springs-eternal seed that next time we'll have some nonstop action like we see on the videos. I hope so, but I'm still bringing my large thermos.

39
SNOW GEESE AND POLAR BEARS
COLLISION COURSE?

For several years I took my ornithology class to the Dakotas to witness the spectacle of the spring Snow Goose migration. Huge, noisy flocks of migrating geese stretching from horizon to horizon provided an experience we will all remember for a lifetime. The geese would fly right over parked vans and land a short distance away in a field, providing amazing glimpses of large flocks. It was hard to focus on a single bird, with thousands of birds milling about (incidentally, a good reason for flocking behavior, as the same probably happens to predators). We often estimated that we observed at least a million birds during a weekend field trip.

A main reason for this spectacle was the dramatic increase in the population of Snow Geese. Favorable conditions on the wintering grounds led to high overwinter survival, and females could arrive at their Arctic nesting sites in very good condition, necessary for producing a clutch of eggs. This resulted in a huge population explosion. Unfortunately, a consequence of this population increase is that they have been destroying the Arctic tundra nesting grounds. This occurs because of the way they feed: pulling plants out of the ground rather than grazing them (so that they can grow again). This alteration of the environment affects other species and led to the authorization at state and federal levels of a spring hunt, with reduced restrictions and high bag limits. Although I've heard them called "sky carp" or "tundra maggots," they are very tasty in the spring after having fed in southern grain fields all winter.

Our first few trips to the Yankton area of South Dakota for the spring Snow Goose hunt were a blast. We got a few birds over decoys and did some ditch sneaks. We noticed that compared to prehunt times, the behavior of the geese had changed. They now fly high and drop almost straight into the middle of a field, skillfully avoiding a parked car or person on the road like the plague. After trying a few what-the-heck shots and wasting some expensive heavy-shot ammo, I trained my range finder on some overhead flocks and determined they were at least 150 yards up. They had smartened up indeed and effectively flew out of range of shotguns.

Our hunts ended when in our last effort, after chasing spooky flocks and encountering a large number of hunters, our decoys were stolen from a field during the night. Sort of spoiled our enthusiasm, and we haven't returned.

Whether the spring hunt has directly reduced Snow Goose population numbers isn't clear, although it may have had an indirect effect. By chasing migrating flocks around, hunters have interfered with the ability of females to stay in one field and gather enough food to produce consistently large and successful clutches. And there's a new sheriff in town—polar bears!

Polar bears typically feed on ringed seal pups on the pack ice of the Arctic oceans (including Hudson Bay) during the spring. With the earlier and earlier disappearance of a large percentage of this habitat (see graph), caused by global warming, polar bears' ability to catch young seals has also disappeared. However, recent observations have shown that the bears have a trick or two up their coats. The bears are starting to come ashore earlier than normal because of the early disappearance of sea ice, where they find colonies of Snow Geese sitting on large clutches of eggs. Snow goose omelets anyone? The birds, not having had to cope with a predator like a polar bear, make no attempt to hide their eggs or put their nests in inaccessible places (there are few such places on

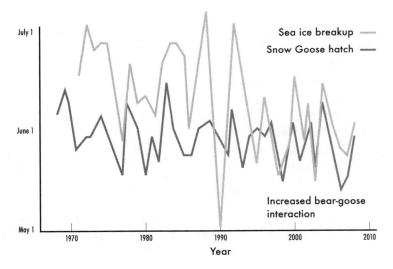

Timing of sea ice breakup (upper line) and Snow Goose hatch (lower line) in western Hudson Bay (data from Rockwell and colleagues). The timing of the two becomes more and more intertwined toward the present, resulting in polar bears coming ashore earlier than normal and preying on Snow Goose eggs.

the flat tundra). The bears can in short order (pun intended) wipe out a nesting colony of geese.

Will polar bear predation compensate for the lack of human hunters' success in reducing goose numbers? Robert Rockwell from the American Museum of Natural History in New York City and his colleagues from the museum and Utah State University made some theoretical models that predict the fate of bears, Snow Geese, and their interactions along the Cape Churchill Peninsula of western Hudson Bay. They predict that the bears will reduce the Snow Goose population but not wipe out the geese. The reason is that the overlap between the nesting season and ice breakup is not perfect, as the graph shows. Thus, some years the bears will not get to the Snow Goose nesting grounds in time to feast on eggs, but other years they will. In the years bears don't make it to the nesting grounds in time for breakfast, the geese will nest successfully, keeping goose populations going.

Although Rockwell and colleagues' study does not predict the total annihilation of the nesting Snow Geese along Hudson Bay, the decline could be substantial. For example, assuming a starting population of fifty thousand nesting pairs, their models predict that in twenty-five years there will be around five thousand nests, a reduction of 90 percent. If this were true everywhere, the Snow Goose numbers would be reduced far below the targets set by U.S. and Canadian agencies. Of course, the reduction might not happen, because Snow Geese nest in other areas, where there are fewer polar bears, although the distribution of bears and Snow Geese is incompletely known. Incidentally, polar bears in Europe are also forced inland by early ice melt and are preying on other species of geese, such as the Barnacle Goose. So, we can't claim that our bears are particularly innovative.

Many factors can complicate the outcome for Snow Geese. In some years, the bears may arrive only in time to eat eggs from late-nesting geese. Snow Geese live many years, and if they lose their eggs in one year, they may be successful the next. Alternatively, polar bears may move ashore early even if the ice does not disappear, if they develop a taste for eggs. In fact, Rockwell remarked to me, "I was gratified this spring to find a fat subadult male standing amidst thirty to forty empty Snow Goose nest bowls—he has obviously read our work, came ashore early, and was feasting!" He also mentioned that in some areas, seal-hunting bears come ashore for an "egg break" and then go back to hunting seals.

The unfolding saga of the consequences of global warming is well illustrated by the interactions between polar bears and Snow Geese. Even the extreme global-warming doubter has trouble explaining why sea ice is melting earlier and earlier, as shown in the graph (not the same as explaining what's causing global warming, be it humans or natural factors). This is pretty basic stuff—ice melts in warm water. Clearly, the goose and bear are entering into a new era of interactions. Perhaps Snow Geese should nest earlier, although they might already be nesting as early as possible.

Polar bears may not survive if all they have to eat are Snow Goose eggs. Some scientists in fact predict the extinction of the polar bear. Rockwell recalled a quote from an old Inuk hunter, who said something like, "Things are changing, and the bears will change too. Some will die, and some will be skinnier. But just like the Inuit, they will still be around a long time from now." Rockwell says he's betting on the bears to survive. Time will tell, but I hope he's right.

40

SPECIES CONSERVATION AT THE STATE LEVEL
A FISH-EYE VIEW

Most people have heard of the U.S. Endangered Species Act (ESA), federal legislation passed by Congress in 1973, which extends protection to species, subspecies, and distinct population segments (of vertebrates only). Recognizing that our natural environment provides "esthetic, ecological, educational, recreational, and scientific value to our Nation and its people," the ESA is designed to "protect and recover imperiled species and the eco-systems upon which they depend." A species can be designated as "endangered" or "threatened" depending on whether its entire range or only a portion is at risk. There are just under fourteen hundred listings in the United States.

Perhaps less well known is the fact that legal protection can be extended to species at the state level. As an example, Minnesota defines "endangered" as a species threatened with extinction throughout all or a significant portion of its range within Minnesota. "Threatened" means a species is likely to become endangered within the foreseeable future throughout all or a significant portion of its range within the state. The Department of Natural Resources' (DNR) web page states that "a person may not take, import, transport, or sell any portion of an endangered or threatened species." However, there are some exceptions, such as if an herbicide applied to agricultural crops accidentally kills an endangered or threatened plant. A species is listed as "special concern" if it is extremely uncommon in Minnesota or has unique or highly specific habitat requirements and deserves careful monitoring of its status. Special concern species are not protected by the same rules that apply to threatened or endangered species, although habitat modifications or regulations can be proposed to protect the species from current threats.

The Minnesota DNR has been very proactive in surveying and evaluating the population status of many species in the state. They just proposed over three hundred changes to the status of species on the list, which includes birds, mammals, amphibians and reptiles, fish, mollusks, caddis flies, jumping spiders, vascular plants, mosses and liverworts, fungi, tiger beetles, moths and butterflies, leafhoppers, and dragonflies. Much of our information on these groups has come from the Minnesota Biological Survey. In a sense, the list of threatened species provides an index to the environmental health of the state.

I went through each list to see how Minnesota is doing, and counted the number of species that were given a higher rating of concern (e.g., moved from threatened to endangered, or from none to special concern) and the number whose threat level was downgraded. Unfortunately, for most species groups, upgraded threat levels outnumbered downgraded ones 251 to 47. Granted, in most cases the increase in threat level has resulted from better information about where species are and are not, especially for the relatively unknown groups, and not a recently increased threat per se. I decided to look at fish in more detail.

Twenty-one fish species had an upgraded threat level compared to none that were deemed to be doing well enough to be downgraded. Thirteen species were listed (as special concern) for the first time, four went from special concern to threatened, and four from special concern to endangered. In some cases, pretty good records exist to support the change, in others the data are less clear. For example, the American eel, found mainly in the Mississippi, Minnesota, and St. Croix River systems, was listed because very few have been captured relative to previous times, although there are no quantitative data because eels are hard to survey scientifically.

The skipjack herring was once found commonly in rivers as far north as Minneapolis, Taylors Falls, and Big Stone Lake. Skipjacks bred in Lake Pepin. However, they undertake an annual northward migration to spawn, and dams, such as the Keokuk in southeastern Iowa (built in 1913), blocked this essential part of their reproduction. The species was unreported for decades in Minnesota. In high water years of 1986 and 1993, skipjacks were again collected in Lake Pepin for the first time since 1928, and the species was listed as special concern in 1996. Since then, the species has rarely been found even though the DNR conducted extensive surveys, and importantly no juveniles were collected. Thus, the DNR concluded that "while further research into the species' life history and ecological requirements is needed, it is needed and reasonable to reclassify the Skipjack Herring as

Endangered in Minnesota." They also noted that if we are to help reestablish this species, we need to provide some ways for them to circumnavigate the many locks and dams that block their historic routes.

Most of the fish on our state list are not game species. One exception is the northern longear sunfish, which was added as a species of special concern. The species is restricted to northern Minnesota lakes with very high water quality, firm substrates, long shorelines with emergent vegetation, and extensive shallows of submerged vegetation, where males defend in-shore nests. The species is vulnerable to shoreline development, removal of riparian vegetation, and construction of swimming beaches. The DNR surveyed 1 river and 119 lakes in northern Minnesota, finding these sunfish in 11 lakes previously known to contain them and 12 new localities. However, because its distribution is scattered and it is sensitive to the kinds of lakeshore development that reduces water quality, the DNR concluded that "it is needed and reasonable to classify the Northern Longear Sunfish as a species of Special Concern at this time." So if you catch one, you should let it go.

Many of the species on the list are admittedly at the margins of their range. This has been a source of contention in threatened and endangered species management for some time. Biologists expect species to be less common at the limits of their habitat requirements; that is, they are not equally common at the periphery of their range as they are in the center. Thus, to list or not to list peripheral populations is the question. In part, it should depend on whether the species is doing well in the main part of its range; if not, perhaps listing is warranted. I am skeptical about the value of listing species at the limits of their ranges, especially if they are doing well in the central portions. On the other hand, given the range changes expected with global warming, perhaps species that are only marginally found in Minnesota today will find Minnesota the new center of their range in the future, and we should attempt to save local favorable conditions.

Why should we care if a few fish species, especially at the

margins of their ranges, are in trouble in Minnesota? I think we should liken the DNR list to taking the temperature of a sick person. Maybe she has only a slight temperature at the moment, and we can skip a trip to urgent care, but it has the potential to rise greatly. We have altered our waters in a way that species at their range limits are doing less well. It is not a stretch to think that continued alteration of our aquatic environments will start to impact the common species. Many of these nongame species are integral parts of the food web upon which our game species depend. Now that hits home to me, and I applaud local natural resources agencies for their vigilance. We cannot realistically expect the U.S. Fish and Wildlife Service to monitor all species everywhere, and it's to the state's well-being for them to keep track.

ANIMALS AND US

41

RECONSIDER YOUR WALK WITH FIDO?

Most people have heard of the negative effects that house cats have on our native wildlife. Cats on the loose kill about a million birds a day, and they kill an even larger number of native rodents (wonder why we have fewer raptors?). But what about man's best friend?

Just when you thought we had identified almost every bad thing we do to the environment, here's another one. People walking dogs (on leashes!) reduces the numbers of bird species and individual birds present, at least along the trails often frequented. That's what Peter Banks and Jessica Bryant wrote in a scientific paper titled "Four-Legged Friend or Foe? Dog Walking Displaces Native Birds from Natural Areas," published in the journal *Biology Letters*.

OK, let's say that you were interested in whether walking your dog reduces the number of local birds. How would you go about finding out? You can't just walk around your backyard or ask a few friends. To make the study scientifically sound, you have to jump through quite a few hoops.

Banks and Bryant surveyed ninety different trails in urban-fringe woodlands twenty miles north of Sydney, Australia. They identified forty-five trails where leashed dogs were walked, and another forty-five trails where dog walking was prohibited. They had two kinds of experiments: one in which a person and dog walked the trail, and one in which only a person walked the trail. In each case, an observer followed right behind and used a standard method to census birds within 250 meters of each side of the trail. To make their study as realistic as possible, they used various breeds and sizes of dogs, and people of various heights, and censused both mornings and afternoons. In addition, just an observer was present on some walks (although as the authors admit, this is not very different from the walker alone) to get a "baseline" measure of bird diversity.

Now at this juncture, I was wondering, what's the point, anyway? I'm sure that birds fly away from a dog (reacting less so to one on a leash), but they must get habituated to leashed dogs in the same way the squirrels in my backyard taunt my drahthaar by staying just close enough to a tree to escape his endless, but hopeful, charges. But some people are seriously concerned that native birds, at least in Australia, see dogs as potential predators, and if they see enough of them, they'll choose to live elsewhere, making these "natural areas" less birdy than they'd otherwise be. Not to mention reducing their value as a natural reserve.

The researchers found that in both kinds of areas, "dogs allowed" and "dogs prohibited," the walker with dog resulted in the fewest number of bird species detected. As one might expect, the solitary walker had an intermediate effect. Why the results were different for the walker plus observer versus the observer alone is somewhat puzzling, but apparently two people on the trail make more of a disturbance.

The results of this study make pretty clear that walking your dog disturbs native birds. In the "dogs prohibited" area, the number of species recorded went from 4.9 (just observer) to 2.9 (walker plus dog); that is, in an area where dogs are normally prohibited, walking a dog had a relatively big effect. Also, as you would expect, ground-dwelling birds were more affected than those living in trees. I can easily understand how walking my pointing dogs would cause birds to avoid the trail while we were there. I have more difficultly seeing why it would have a long-term effect. Don't the birds just return after we pass? Haven't they figured it out? The dog is gone! The real question is whether more birds were in the "dogs prohibited" areas when there were no dogs present. That to me is the big question. Comparing the "observer alone" results for the two areas, we see that more species were detected on average in the "dogs allowed" area (5.3) than in the "dogs prohibited" area (4.9). However, because Banks and Bryant did so many surveys, they found that this difference was not statistically significant. So, each type of area had on average the same number of bird species and individual birds. Good news for dog walkers, given no long-term effects?

We can interpret these results in a couple of ways. On the one hand, I'd have to say that because there were no long-term effects, who cares—walk Fido as much as you want. On the other hand, birds are affected for the short term, and if this disturbance occurs in the nesting season, the impacts could be significant. First, repeatedly flushing birds from nests allows eggs or young to cool or be fed less often, and second, birds leaving and returning more often gives nest predators a greater heads-up as to nest location. So, maybe disallowing dogs on trails in the nesting season would be advisable (just as we prohibit hunting dog field training during our nesting season). But birds may avoid nesting close to trails in "dogs allowed" areas anyway.

Millions of people walk their dogs in natural areas. In fact, in some European cities it is a legal requirement to walk your dog outside. Although Banks and Bryant's study contains some points to quibble with, it will be used by managers of conservation-sensitive areas as the reason to prohibit walking dogs. In fact, here is how Banks and Bryant ended their paper: "Our results therefore support the long-term prohibition of dog walking from sensitive conservation areas." And their study has some more wide-ranging implications too. For example, their results could be used as an argument to keep people from walking trails, owing to the reduction in bird species and numbers of individuals that resulted from the walker alone.

Once again we are reminded of our impact on our natural world. Now, we certainly are, or at least once were, members of that natural world. But we may have sufficiently altered the natural order of things that our impacts are wide-ranging as well as subtle. Maybe we should consider walking our dogs in areas deemed less ecologically sensitive; I don't think my dogs would object to a walk in a less-than-pristine area. They might appreciate more dog company.

42

LOON HUNTING
A BYGONE TRADITION

On 6 May 1950, eleven federal and state wildlife enforcement officers staged a long planned, coordinated raid on Shackleford Banks, Carteret County, North Carolina, apprehending nearly 100 hunters, of whom 78 were formally charged with illegally shooting loons (Anonymous 1950) in violation of the Migratory Bird Treaty Act of 1918.

So begins a fascinating article on the hunting of Common Loons written by Storrs Olson, Horace Loftin, and Steve Goodwin in the December 2010 issue of the *Wilson Journal of Ornithology*. I had never heard of loon hunting, and what I learned was some astounding early history of Minnesota's state bird and its past perils on its East Coast wintering grounds.

The Shackleford Banks is a narrow east–west barrier island that separates the open Atlantic on its south shore from Harkers Island to the north, with Back Sound in between. Shackleford Banks was settled by the early 1700s by people who made their living catching whales, mullet, and dolphins. The area was devastated by the San Ciriaco hurricane in 1899, resulting in people moving inland, many to Harkers Island. The banks were mostly uninhabited and were used for grazing cattle; they became part of the Cape Lookout National Seashore in 1966.

Loons are fairly large birds, with males averaging sixteen

pounds and females ten pounds. They were never a common species in the bags of waterfowlers or market hunters, because they are hard to hunt and have a reputation as being too "fishy" to eat. They do not respond well to decoys. Although in winter they molt their flight feathers offshore and are flightless, pursuing them with boats is difficult because, as most of us know, when they dive, it is hard to predict where they'll come up, and they often ride low in the water, making them a difficult target. As Minnesotans know, they require a running start across the water's surface to become airborne. When they are migrating, they often fly very high and well offshore, out of shotgun range. And although their calls have nothing to do with this article, close your eyes, and recall their eerie vocalizations, which fascinate us during their summer residency on our lakes.

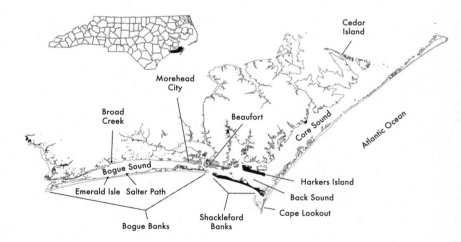

Before dawn, hunters would cross in boats from Harkers Island across Back Sound to the northern shores of the Shackleford Banks. The hunters moved to the outer shore and often lined the entire length of the Shackleford Banks to shoot passing migrating loons in spring. Much of the North Carolina Outer Banks region where these once traditional, but long-since illegal, loon hunts occurred is now designated as National Seashore.

If loons are hard to hunt and taste fishy, why was there a tradition of loon hunting on Shackleford Banks? There was a flaw in the loon's migration strategy. Along Shackleford Banks loons become common in spring and take off to the north from just offshore, bringing them low over the banks and within shotgun range. The original Shacklefordians took advantage of this to put meat, albeit loon, on their tables. Harkers Island became the focal point for this activity, although loons were killed in large numbers in other areas. By the mid-1800s loon hunting was well established.

Generally in April and May, hunters would depart Harkers Island before dawn, especially with a northeast wind, and beach their boats on the north (protected) shore of Shackleford Banks. They walked southwards across the dunes to await the first flights of loons, which usually occurred at first light. Hunters usually didn't require much concealment, although during World War II

Harkers Island resident Otis C. Willis (born November 1918; died June 1996) with the results of a loon hunt in 1943.

they sometimes hid behind antisubmarine buoys that had broken loose and beached. At times there were so many hunters that they were either stacked up behind one another or were spread out just out of shotgun range of each other; there was often no place a loon could fly over without being in range of a gun. One resident of Harkers Island recalled that "shooting began so regularly at daybreak that it was as good an alarm clock as you could ask for" and that at one point during World War II "the shooting sounded like an invasion."

The loons were taken home without being field dressed. This doesn't seem particularly odd, but one hunter might shoot over fifteen loons, which at ten-plus pounds each amounted to quite a load to drag back to the boats.

I figured that loon hunting was more about shooting sport than food. Instead of sporting clays like we shoot today, they had "sporting loons." However, the residents vigorously defended loons as good table fare. They even claimed that they preferred them to ducks and geese and that the fishy taste was "barely discernible." One cookbook refers to loons as "Harkers Island turkey." Olson and colleagues noted that "for 19th century residents of Shackleford Banks who had survived a winter on salt fish, root crops, and such grits and flour as they may have received in trade for mullet and whale oil in the previous months, a loon represented a sizeable chunk of fresh meat as well as a welcome change in diet." A bird not consumed by the hunter might bring fifty cents on Harkers Island.

A second use of loons was to make the bones into fishing lures, which were used to catch bluefish and Spanish mackerel. Bones were sold to local hardware stores for ten cents. The bones were bleached and slid over a hook, and the white coloration was apparently attractive to fish.

Killing loons became illegal in 1918 with the passing of the Migratory Bird Treaty Act. The hunting continued in that local area for quite a while, and during World War II, people had other things to worry about. But the hunt became too popular, and the locals were joined by others from some distances away. This was too much for law enforcement to ignore, and in the 1940s an undercover agent settled locally and began documenting the loon hunts and planning the raid mentioned above.

It was estimated that the hunters had killed two hundred loons on the morning of the big raid. Hunters who were apprehended were fined $25 each, which is about $325 in today's dollars. The raid was deemed a big success for law enforcement as it basically put an end to the illegal loon hunts. Shortly after the big raid of May 1950, residents reacted negatively. According to Olson and colleagues, S. Guthrie wrote in the May 12, 1950, *Carteret County News-Times* that laws that overregulated hunting had come into effect because of "fancy" "upstate" sportsmen, whereas the Carteret County loon shooters participated in the "only exciting sport left untouched by the law. Their father had done it, their grandfather before, even before their forefather had moved from the outer banks. You weren't a man until you had shot a loon."

I don't bear a grudge against the loon hunters, as I think you have to understand the context in which they participated. It is thought that a few loons are still killed today and secretly end up in local stews. It's hard for me, not having been part of the local culture at that time and place, to pass judgment from afar. I'm glad to have read about the tradition and glad it's ended. Nothing adds more to my northern Minnesota experience than the sounds of loons.

43

MARKET HUNTING AND THE DEMISE OF THE ESKIMO CURLEW

At this season [during fall migration] they are considered by epicures the finest eating of any of our birds, and consequently they are watched for and sought after by sportsmen with great perseverance during the very short period that they are expected to pass along this coast during their migration southward. (George H. MacKay in the 1892 *Auk* [published by the American Ornithologists' Union])

It is not hard to imagine that if this quote referred to your favorite species, its long-term survival would be in doubt. Indeed, the species is the Eskimo Curlew, and although it was once abundant, today most consider it extinct. The Eskimo Curlew is about thirteen inches in length and weighs a pound at most. It is brownish and has a decurved bill.

Still, if it was abundant, what happened? One likely culprit was unregulated "market hunting," as opposed to overhunting by sportsmen. My friend and colleague Gary Graves from the Smithsonian Institution recently summarized the past market hunting of these shorebirds by digging in old journals. A sad tale and a cogent lesson emerges from his 2010 paper in the journal *Waterbirds*.

The Eskimo Curlew is/was an Arctic breeder and wintered in southern South America, undertaking the impressive migratory trip twice per year. In fall, they migrated along the Atlantic coast from Labrador to Long Island, then out over the open Atlantic to South America. "Vast numbers" of migrating curlews were seen in autumn. In spring, they migrated up the Mississippi River valley. Many birds that migrate south along the coast in fall migrate inland in spring because offshore and unpredictable spring hurricanes are frequent enough to be devastating to migrating birds. Old reports mention spring skies filled with curlews as they migrated through the prairie states and provinces. One report describes a single flock feeding in Nebraska that covered forty to fifty acres, where they were feeding on grasshoppers and other insects. True, they were not on a par with the estimated billions of Passenger Pigeons, but they must have been quite a sight.

Much of Graves's analysis came from the writings of George H. MacKay (1843–1937), "an ardent sportsman and conservationist." MacKay published many short scientific accounts, which today can be viewed freely online, but Graves used information from "a shooting journal which contained data from Nantucket Island that never appeared in his earlier papers or in later compilations." In this journal MacKay recorded the numbers of Eskimo Curlews (and American Golden-Plover) observed and shot, the names of those who hunted with him, and the dates, times, and places they hunted. The journal covers the period from October 1, 1865, to August 16, 1922; for some unknown reason, MacKay did not hunt for twenty-four years between August 29, 1897, and October 7, 1921. He was in the shipping business and apparently a man of means, summering on Nantucket, and usually returning to Boston after the last shorebirds had passed south, but he "occasionally lingered on Nantucket until October." Nice work if you can get it.

During his outings, MacKay and his partners observed 650 curlews on thirty-eight days and killed 87 birds in total over twenty-two hunting seasons, achieving a high count of 16 bagged on August 27, 1877. MacKay's journal reveals problems with curlew populations. Graves determined that "Eskimo Curlews were observed on 23% of hunting days logged during the flight period in the 1870s, 19% of days in the 1880s, but only 4% of days in the 1890s." MacKay observed only 5 curlews in seventy-seven hunting days during the 1890s. This decrease was also reflected in the number harvested. The last few years they shot 2 or fewer per year, and he shot the last one he observed on August 21, 1893.

MacKay began hunting American Golden-Plover after the last major curlew flight in 1881. They too were deemed wonderful table fare. They shot a total of 927 plover, and on some days in 1880, for example, more than 30 were taken in a day. This was legal, by the way. However, like the curlew, annual bag totals of plover declined steeply during the later decades of the nineteenth century. Again, market hunters played a role.

MacKay was aware of, and his journal documented, the steep decrease in numbers of curlews and plover. In one of his scientific publications he criticized market hunters for causing these drastic declines. This judgment might seem hypocritical, given that he apparently saw nothing wrong in hunting the birds himself with his friends in the fall. Some detective work on his part justifies his stance.

MacKay visited "game dealers" in Boston, places where curlews, plover, and other wild game were sold to the public. In the mid-1880s these dealers were receiving large shipments of Eskimo Curlews and American Golden-Plover that had been taken in

Nebraska, Missouri, and Texas during spring migration. The birds were probably shipped to eastern markets in train cars packed with ice or cooled by a primitive refrigeration system.

In 1890 two dealers received twenty barrels of birds, each of which contained one-third American Golden-Plover and two-thirds Bartramian Sandpipers (now called Upland Sandpipers), eight barrels of just Eskimo Curlews, and twelve barrels each containing Eskimo Curlew and American Golden-Plover. There were twenty-five dozen (300) Eskimo Curlews and sixty dozen (720) American Golden-Plover in a barrel. My guess is that there were seventy-five dozen Upland Sandpipers in a barrel.

These game dealers received about 4,000 Eskimo Curlews and 9,360 American Golden-Plover in a single season. MacKay could only speculate about how many were killed if other large cities were similarly supplied. Indeed, these numbers were a major take for each of these species. Incidentally, Upland Sandpipers are also not doing well.

Even if many other hunters like MacKay that hunted curlews and plover during migration, the numbers they took likely paled in comparison to what the market hunters were taking. Plus, in spring the birds they shot were likely destined to breed, whereas in the fall many immatures were taken, many of which might not have made it back to breed. As population biologists understand, a large number of birds do not return from their southward migratory journey, due to natural factors such as bad weather, predation, and lack of sufficient food.

We need to remember that market hunting occurred in an era when people didn't imagine that we could bring down entire species with indiscriminate hunting. And the Eskimo Curlew was targeted because it was so common, as was the Passenger Pigeon and American bison. Market hunting of birds stopped with the passage of the Migratory Bird Treaty Act in 1918, although not many Eskimo Curlews were probably left after 1900. Public opinion might have stopped market hunting even without legislation, but it was probably too late.

Unfortunately, unlike many other species that were market hunted, such as ducks, egrets, and herons, the Eskimo Curlew was unable to rebound because too much of its prairie habitat had been converted to agriculture. Incidentally, the last clear sighting of an Eskimo Curlew in the United States was in 1962 in Texas, and one was shot in Barbados in 1963. Unconfirmed sightings trickle in sporadically. So, there is a slim chance that some curlews survive and have escaped detection over the past fifty years. Unlikely, though. From Grave's reconstruction of the last good days for the Eskimo Curlew, we have a valuable reminder that we can indeed push a species to the brink of extinction with unregulated hunting.

44

THE ETHICS OF BAITING AND HIGH-FENCE RANCH HUNTING
A PERENNIAL DEBATE

What do baiting deer and hunting at high-fence ranches have in common? Ethics. Few topics are as guaranteed to start an argument as different hunters' perceptions of what is ethical. For example, many factions have dug in their heels and deemed baiting to be akin to sleeping with the devil himself. There are some good reasons for this opinion, both ethical and otherwise. If a white-tailed deer in the latter stages of a Chronic Wasting Disease infection shares a corn pile with uninfected deer, prions (the infectious agent) could be transmitted via saliva to healthy deer, and a CWD epidemic could result, at least in theory.

The remaining concerns about baiting are largely ethical in nature. Some are concerned that bait piles change deer movement patterns, but so do snowmobile/ATV/hiking/biking trails, roads, subdivisions, new houses, agricultural fields, gardens, nature centers, fences, and so on. Many hunters bring up the possibility that antihunters will have an even more negative view of hunting if deer are being shot over bait. It's true that hunters do not need more scrutiny and negative publicity.

Not all "bait piles" are the same. A garden, a food plot, a bird feeder, an agricultural field, or a compost heap can legally attract deer away from the natural vegetation upon which they should be feeding, where they are considered an ethical target to many. In contrast, some people think that food plots are glorified bait piles (my food plot has corn, apple trees, pumpkins, clover, and ryegrass). This is obviously a slippery slope. We bait bears in most eastern states, but not deer, as it is very difficult to stalk bears, and many states use hunting to control bear numbers. Some states allow both deer and bear to be baited, some neither. If local game laws reflect what the majority of hunters feel is ethical, obviously hunters (or legislators) in different regions of the country disagree about what constitutes ethical hunting. In principle, sentiments against baiting trace to the well-known North American model of wildlife conservation that provides one definition of fair chase, which is inconsistent with baiting. The issue is that the definition of fair chase amounts to a subjective judgment about what is ethical, a perennial problem for human beings.

Another sure-to-generate-debate topic is hunting at high-fence ranches. In states such as Texas, many private fenced ranches offer the opportunity to hunt native deer as well as a host of exotic species, such as feral hog, axis deer, sika deer, red deer, fallow deer, aoudad, blackbuck, goats, and rams of many flavors. You can even save the money you would need for an African trip and shoot a zebra, wildebeest, oryx, or water buffalo in Texas! And to compound our ethical debate, most of these places use bait to lure game into bow or gun range. Shooting baited animals in a high-fence-enclosed ranch sounds like a publicity nightmare for hunters, and surely there must be a consensus that this is not ethical. Obviously, given the large number of successful high-fence hunting operations, many consider it "ethical enough" to partake. So do I.

I have visited a high-fence ranch in Texas with my two sons starting when they were ten. The first question I get is usually, "What is it like, and isn't it like shooting animals in a pen?" Short answer, no. All the animals are "free ranging" but confined in a high-fenced area, the size of which depends on the ranch (usually 150 acres and up). Or, conversely, if you're against this activity, you would say that by definition it is unethical hunting because "the animals are trapped in an enclosure from which they cannot escape." This gives the illusion that their harvest has little to do with the hunter's skill, and that you walk around after them until they think you're going to feed them, and then shoot them. Of course on the high-fence ranch where my son and I bow hunted in Africa, the fenced area was 10,000 acres, and two female rhinos with calves wandered the property. So, in this case, I think the animals are free roaming, and if someone wanted to walk around with a bow, I would have to admire their lack of common sense and hope they could outrun an enraged cow rhino.

Many U.S. hunting ranches have a network of dirt roads or trails. After you are dropped off in your hunting spot, one of the ranch personnel drives the roads, laying down a thin layer of corn. Animals typically come to the road and feed on the sparsely distributed kernels of corn. You usually hear the animals first, especially the hogs. In theory they will feed past your shooting window, giving you a shot opportunity.

So it should be easy to close the deal—just about like shooting them from the corral fence? A perfectly unethical setup? This is an oversimplification. The animals are typically wary, especially hogs that have been hunted sometimes every other day all year, way in excess of natural hunting pressure. Enough of them know where most of the hunters will be stationed. So at even a hint of movement, smell, or noise, the whole group vanishes, and you are embarrassed that you were fooled by these "tame" animals. It's easy to underestimate these hogs and exotics (except for rams and goats, which are frankly pretty dumb). We wear camouflage and face masks, reduce human scent as much as possible, hide to the best of our abilities, and are as quiet as possible. I'm pretty sure that some deer hunters don't take all the precautions we do. Even a skilled bow hunter will find these hunts challenging.

The "hunts" are not a guaranteed deal. Most times I sit without shooting anything. For example, one morning I was in a well-constructed brush blind eighteen yards from the road and saw nothing. The corn sat untouched. Recent rains provided ample new vegetation growth, and the animals forgot that they were obliged to visit the road. That afternoon, I was in the same ground blind, and about an hour before dark I heard pigs foraging on the road, headed my way. They were active and noisy, and from a peek hole in the brush blind I could see them almost leapfrogging over each other. I was concealed by a thick tangle of brush and drew back my bow when the lead pigs (usually the smaller ones, which the hogs seemed to have figured out are safer, and hence the scouts) were about fifteen yards from my shooting lane. I made no sound, they could not see me, and the wind was in my favor. After a couple of seconds I realized something was wrong— no noise was coming from the road. They had veered off about twenty yards from my opening and headed off into the brush—all of them. Hogs, one; hunter, zero.

I was really frustrated, as I'm sure they didn't hear, see, or smell me. Maybe they detected that someone had recently walked into the blind from the road. However, all was not lost. In about five minutes, another, smaller group came from the same direction. Same story: bow pulled back, heart pounding, hogs coming, hogs gone, no explanation. This is supposed to be easy?

My two sons were also hunting. That day, they got one hog between them, a nice boar that my older son shot that evening. A short tracking job was made easy by aid of a fearsome hog-tracking dog named Pickles. Pickles is supposed to find the downed animal and bark. Instead, Pickles usually finds it, takes a sniff, and returns to the handler, or feeling satisfied, goes off for a nap. So, the plan was to "follow the Pickle." Another of the dogs, a diminutive beagle-like dog, was named Gator.

So much for easy. I had had no shots the first three hunts. I harvested a blackbuck doe the night of the second day and another the morning of the third. I didn't get another shot opportunity until ten minutes before dark on the last night (eighth hunt), when just two hogs came down the road. I drew back, and the first hog came up to my shooting lane and then sprinted across the opening, although I am sure he could not see, hear, or smell me. He just knew there was a stand there and wasn't taking any chances. Still at full draw, I could see the second one coming but decided to twist around and shoot the first one, which was now broadside at twenty-five yards. I watched my arrow skip under the hog and vanish harmlessly into the brush (we recovered it later, and it was a clean miss). "Hog fever" caused me to botch a "gimme." I ended my eight sits with two blackbuck does and a missed hog. My sons each ended up with two animals each. Final tally was six animals for twenty-four hunts; in other words, we had a one-in-four chance of harvesting an animal on any given morning or afternoon hunt. This result is better than my typical whitetail hunts, in which it's more like one in a dozen. But this kind of hunting is not a sure thing and nothing like sitting on the corral fence.

Another reason I like these hunts is the opportunity to spend time with my sons and, more importantly, give them practice at bow hunting. So yes, I consider this very serious "practice hunting." My sons have learned about when to pull back, being quiet, playing the wind, making ethical shots (they have passed many marginal shots), waiting the right length of time to track, and so on. They have learned a lot about themselves. I have learned a lot about them. We consider our hunts quality experiences. I do not apologize for taking my sons bow hunting where bait is used. I do not think I have tainted them. They understand the circumstances and that they will usually see fewer "wild" animals.

Some of our other experiences transcend hunting. Once we picked up my older son, then fifteen years old, and learned that he had shot a hog. We were all excited, as he had seen it go down and had already recovered it. As we were riding back to the bunkhouse, he asked me what I thought was a rhetorical question: "Dad, can rattlesnakes hear?" Well, I said they could sense vibration (I've since learned that they can indeed "hear"), but I had a feeling the question was more personal than academic. Sure enough, he told me that after making the shot and seeing where he thought the hog went down, he was going to look for it. As he was about to climb down from his stand, he heard something in the leaves and saw a six-foot-long rattler below him about ten yards away. He yelled at it, and it kept coming toward his stand. Then he stomped on the stand's ladder, and it stopped. Now the story got fuzzy here, but for some reason he lost sight of it, but he thought it had left the immediate vicinity. So he climbed down from his stand and found his hog. My reaction, expletives deleted, was *why* would you get out of your stand with a snake of that size in the immediate vicinity. His response: "You know, Dad, it's not like it's going to chase me down or something." He wanted to find his hog in the last light and bring it back to the road all by himself. As a snake-a-phobic, I was "rattled," but I guess he had a point.

On another trip, my younger son—who was ten at the time—was on his first bow hunt to Texas. At the end of an evening's hunt, we arrived at his ground blind, and he ran out and exclaimed, "Dad! I shot a ram!" He and his mother (his spotter) were pretty sure it went down about fifty yards away. For months prior to the hunt, I had coached him over and over that if he shot an animal, he was to stay in the blind and not risk pushing a wounded animal; we and the dogs would all track it. This is standard for bow hunters; unless you see the animal go down, you wait, find, and examine the arrow and then decide whether to track then or wait.

I didn't, however, anticipate all possible scenarios. He shot the ram thirty minutes into the three-hour hunt on a warm afternoon. About an hour after he shot it, vultures started to circle and land in trees above where he and his mom thought the ram had expired. His mother thought they should get out and scare the vultures away. He dutifully reminded his mother of Dad's rule that "under no circumstances" should you leave the blind. Of course, vultures landing on his kill was not something I had anticipated. Eventually the vultures flew off because my wife waved her hand out a blind window, over my son's objections. His mother rolled her eyes as my son told me how he had followed my instructions to the letter, but his mother had not. I reconsidered my instructions to "stay put" and amended them to include not only the possibility of vultures but a rapid exit if a rattlesnake poked its head under the blind.

In summary, my goal has been partly to describe what hunting is like on one of the high-fence ranches in Texas so that readers can make up their own minds as to whether it's within their realm of ethical. To me, it's exciting and it's ethical. I am not bothered that the animals come to the road to eat corn, any more than that they come to my food plot. These are not wild animals, nor are they tame, and it is hunting to me. I worry about scent and wind, drawing back unseen and unheard, making a clean killing shot, tracking, and so on, just like I do during "real" hunting. There is no guarantee you'll get a shot, and when you do, no one tells the animal to stand broadside and quit moving. Any time the outcome is not guaranteed and it is up to the hunter to be in bow range and make an ethical killing shot, it is hunting to me. However, I freely admit that I don't consider these animals to be trophies, although I might have an axis deer mounted if one ever gets within range to shoot.

Back to ethics. Is shooting animals in an enclosed area over bait ethical? That depends on your point of view. I think we have to realize that what is ethical is in the eye of the hunter—one man's trash is another man's treasure, or something like that. A friend and I visited a ranch in Oklahoma, where he arrowed a really nice Texas Dall ram. He had it mounted, whereas I do these hunts for the meat and don't consider them trophy hunts. We can imagine a hypothetical continuum from hunting a grizzly with a pocket knife, surely ethical if not suicidal, to shooting a deer remotely from a computer, which very few would consider ethical. But in between is a lot of latitude. For example, what about "hunting" released pheasants on a game farm, does that count as hunting? Does shooting a trophy buck at a food plot remove the "fair chase" label? I once wrote an article responding to one titled "Antler Religion" in the *Wildlife Society Bulletin*. The author railed on about how unethical it was to hunt deer in an enclosed space (size not specified). I think that if an activity is legal, like shooting a deer in an enclosure, then you should just choose not to partake if it offends your personal definition of what is ethical. When one group of hunters starts telling others what is ethical, we are on a slippery slope indeed, one destined to do no good for hunting.

I am taken by the subtle distinction between different views of what's ethical. On a bow-hunting website, a well-known writer and editor, a critic of high-fence hunting, praised an outfitter who, on the second morning of making a bow-hunting video, got him within sixteen yards of an elk. The outfitter wrote on their web page: "We spend numerous hours during the off-season locating and tracking elk and mule deer in their natural habitat and continue to track their movement during hunting season to help assure our clients of a true trophy adventure with a high probability of harvest." This writer harvested a wild animal legally, but to me his hunting experience sounds similar to our bow hunts in south Texas, especially the part about the high probability of harvest. The hunt was legal, and I think most would consider it ethical, but some for sure would not. I keep coming back to the viewpoint that if it's legal, it's ethical. If it's not to you, then show your opinion by not supporting it.

45

HUNTERS AND CONSERVATIONISTS AT ODDS OVER SHOOTING SHOREBIRDS

The ornithological community was up in arms recently over the shooting of two large shorebirds, called Whimbrels, on the Caribbean island of Guadeloupe, in French West Indies. Of course the question is, how would ornithologists know this even occurred? The answer is that both birds had been outfitted with satellite transmitters two years ago by scientists at the Center for Conservation Biology at the College of William and Mary (the project also involves collaborations with the Nature Conservancy, the U.S. Fish and Wildlife Service, the Georgia Department of Natural Resources, the Virginia Coastal Zone Management Program, and the Manomet Center for Conservation Sciences).

The island of Guadeloupe has a small but dedicated group of resident shorebird hunters. Each fall they shoot fairly large numbers of shorebirds, including Lesser Yellowlegs, Pectoral Sandpipers, Stilt Sandpipers, Short-billed Dowitchers, Greater Yellowlegs, and American Golden-Plover. These species all occur in Minnesota regularly during spring and/or fall migration.

Shorebirds were once hunted extensively in the United States both by sportsmen and market hunters. They are considered excellent table fare. I described earlier how the take by market hunters, not sportsmen, likely led to the extinction of the Eskimo Curlew, another large shorebird, and to large declines in other species, such as the American Golden-Plover. George MacKay from Nantucket Island kept detailed records of how many shorebirds they harvested each fall, and how this number kept diminishing. At the same time, huge numbers of these same birds were shipped by the barrelful from the center of the country, where they were taken by market hunters during spring migration. This almost certainly led to the extinction of the Eskimo Curlew (and similarly, the Passenger Pigeon).

The two Whimbrels shot were on their southward migration, when they apparently were pushed off course by Tropical
Storm Maria, taking them to Guadeloupe, where they typically
don't stop over in fall. The ornithologists studying the birds had
given them names (Machi and Goshen). This of course heightens
the possibility of their attaching too much emotional significance
to the birds' demise. But what the ornithologists had learned so
far was pretty remarkable. Machi was first caught and banded in
Virginia in August 2009. During the next two years Machi trav-
eled over twenty-seven thousand miles (that alone would qualify
her for a free trip on some airlines), including seven nonstop
flights of more than two thousand miles. These feats were part of
her yearly trips from the breeding grounds near Hudson Bay and
her wintering quarters in São Luís, Brazil.

The satellite track showed that Machi made a long arc around
Tropical Storm Maria, which brought her to her fateful visit to
Guadeloupe. The hunter that shot her noticed the radio antenna
protruding from behind her, and the bands on her legs. He turned
the bird in to a wildlife biologist, who reported it, which alerted
the ornithological community. A few days later, Goshen's satellite
transmitter stopped moving in a swamp in central Guadeloupe.
The wildlife officer drove out to the swamp to ask hunters to stop
hunting out of concern they would kill her too. Although her
remains were apparently not turned in, clearly he was too late.
Only a handful of other birds have transmitters (owing to their
high cost), and this loss was a big blow to the project.

Much more could have been learned from the two Whimbrels, and their deaths were unfortunate. However, shorebird
hunting is legal in Guadeloupe (and a couple of neighboring
islands, such as Martinique). The species is also hunted in Thai-
land. So these hunters were not villains, no more so than someone who shoots a radio-collared bear or wolf. However, a widely
circulated YouTube video of these hunts raises the possibility that
the birds were not being used for food, which violates at least
my ethic that if you shoot it, you eat it. In the video the hunting

almost looked more like practice for the retriever. However, just one hunter was featured, and I have no idea what is behind these hunts in general. After all, large numbers of shorebirds come ashore during the fall, and they have for years, otherwise the tradition wouldn't be in place. One source stated that some hunters shoot between five hundred and one thousand shorebirds each fall (especially after hurricanes and tropical storms push birds ashore). There are apparently no bag limits.

In contrast, on Barbados, hunters have voluntarily agreed not to shoot American Golden-Plover or Red Knots, birds whose populations are in serious decline. It is hoped that the governments of Guadeloupe and Martinique can be persuaded to adopt at least bag limits, put some species on the no-kill list, and even establish refuges, which would benefit many birds and other animals. Perhaps this publicity will serve that end.

I think we should not judge from afar the hunters who killed the two Whimbrels, as we are neither part of the local community nor privy to their traditions. However, apart from snipe and woodcock (rails are not shorebirds), there are no seasons on shorebirds in the United States. This is the case for likely several reasons. First, we proved our ability to reduce numbers of these birds greatly, albeit mostly through market, not sport, hunting. Second, many of the smaller species are hard to identify, and regulation would be a nightmare. Lastly, most of the species that would be good eating, like Marbled Godwits, Hudsonian Godwits, Upland Sandpipers, Willets, Greater Yellowlegs, Long-billed Curlews, and Whimbrels, are uncommon enough that they could probably not withstand hunting pressure.

It is possible to find statements that Whimbrels are among the most common of the large shorebirds. However, this status is relative. The worldwide population is likely less than one hundred thousand individuals, which might not sustain sport hunting on a large scale. I speculate that the bag limits would likely be one or two. In the early 1900s it was reported that hundreds of thousands of Whimbrels passed through South Carolina on the Atlantic coast of the United States during early May. During autumn, they were hunted in the United States during migration. This hunting greatly reduced populations, and although hunting is prohibited, the species has not recovered to prehunting levels. Apparently habitat degradation on the breeding grounds and migration routes will not sustain hundreds of thousands of birds anymore.

I am reluctant to say that because we don't hunt shorebirds in the United States, no one should. And we lack solid data to show that shorebird hunting in Guadeloupe and nearby islands actually affects populations. But clearly if everyone adopted Guadeloupe's position, allowing indiscriminate shooting of Whimbrels, the bird would be in big trouble. So, I conclude that Guadeloupe and Martinique should do the right thing and ban this activity, or put in place and enforce small bag limits, just as we banned shooting of these birds as well as other species, like Common Loons. The fate of the Eskimo Curlew and Passenger Pigeon ought to be poignant enough lessons.

46

A CONVERSATION ABOUT HUNTING IN THE NETHERLANDS

I recently visited friends who live in the small Netherlands country town of Gaast, which lies just south of the dike that divides the Dutch Wadden Sea, about two hours northeast of Amsterdam.

Here on reclaimed land are agricultural fields with lots of grazing sheep. In these fields, my hosts and their colleagues study large shorebirds, such as breeding Black-tailed Godwits, and other birds that use the area as a "refueling site" during migration.

My friends have banded the local birds for years, resulting in a fine database from which they learn about the birds and threats to their existence, which include, not surprisingly, intense agricultural practices.

Of course, agriculture often attracts species, like geese, that overextend their welcome. I climbed a short observation tower, which offered a great view of the landscape, and counted many species of birds that don't live in North America. I noticed many Graylag Geese, which would occasionally fly near my vantage point, sometimes stopping to feed in the agricultural fields.

As this was July, I had already started thinking about the early goose season in Minnesota, and so found myself plotting where I would put my decoys and station myself, if it were the hunting season for graylags. I picked out a location with cut fields on either side of a small channel with some tall grass. I figured that it would be a good ambush spot for geese coming from several directions to check out my decoys.

That evening I was at a reception, and my host, who knew of my addiction to hunting, hooked me up with a local guy who was also a hunter (and spoke English). He had been told about my penchant for bow hunting. Our first topic was a house cat that was just across a fence, hunting native birds and rodents in an adjoining pasture. I mentioned that the cat might be a fair target, and he asked, "With a gun or a bow?" I knew then that we would hit it off.

I asked him if he hunted geese, thinking that if there were as many geese in the hunting season as I had seen earlier that day, a return trip with my sons might be worthwhile. He said he had hunted ducks and geese, so I thought things were looking up. I mentioned that I had seen a place nearby that could be a great spot to set up a decoy spread. Things then, so to speak, turned south.

He said, "We are not allowed to use decoys."

"What?" I replied. "You're kidding, right?"

"No, they're illegal." Not using decoys would certainly change the equation. I thought about my setup spot and figured it might not work for just pass shooting. I asked how hard it was to learn their calls.

"We're not allowed to use calls," my new acquaintance said, "They're illegal, too."

No decoys, no calls—I didn't ask the obvious question, "Then why do you bother having a season?"

Technically they don't, but he said that it was possible to hunt the graylags. Beginning in 1977, you need a hunting license, which you must renew annually with the police. You must be at least eighteen, have passed a hunting test, and prove you have a place to hunt.

Seems reasonable, I thought, but because there are no open public hunting areas in the country (read that again and let it sink in), you are required to have written permission from a landowner (or the manager of a state-owned area) to hunt on an area, which must be at least forty acres in size. Only then do you legally have a place to hunt. Your forty-plus-acre spot must be at least 300 meters wide, and you must be able to draw a circle with a 150-meter radius within it.

You can't hunt on Sundays or holidays, your semiautomatic shotgun can't hold more than two rounds (with steel shot), and you must have insurance in case you injure a third party. You must have a permit to own any sort of gun. If you're not from the Netherlands, you have to be invited by a resident of the country and accompanied in the field by a licensed resident.

Now realizing that I would not be returning for the season, I asked why all the restrictions existed. He said in the past hunting was less regulated, and hunters would get permission for a field but then bring in lots of other hunters, and it would be a "war zone." I had to admit that, actually, that's exactly what I had in mind.

"There were dead birds everywhere. But the people objected to the killing, and so we now have a lot of regulations."

I found out later that the percentage of people in the Netherlands who hunt is 0.2. Although the high level of public opinion against hunting didn't close the season, the situation is not what most of us would find encouraging (especially if no one under eighteen can hunt). One could assume that they're all vegetarians, but that's a stretch—someone kills their animals for them.

As our conversation wound down and dusk settled in, I saw a roe deer in a field headed along a fence row, and my eyes started to narrow, and the intercept computer came on in my head as I calculated where I could lie in wait. He could see my reaction and reminded me, "It's illegal to hunt with a bow in the Netherlands."

I loved my brief visit to the Netherlands and would return in an instant. Great place and wonderful people. Somehow, though, hunting has become so vilified that it doesn't seem worth the effort. It was never clearer that we have a good thing in the United States—I wouldn't want our hunting opportunities to be like theirs.

47

BACK FROM THE DEAD
MOTHER GOOSE GOES TO THE POOR HOUSE, COOKED

> Just because someone is low-income, I don't think we should ask them to eat something that most Canadians wouldn't eat.
>
> —Gail Nyberg, executive director of Canada's largest food bank, refusing to permit Canada Goose meat to be distributed to needy families

Once upon a time, we thought we had lost the giant "race" of the Canada Goose. Habitat loss and overhunting of the big geese resulted in their apparent demise. But in the early 1960s, some representatives of the big goose were found in southeastern Minnesota. It wasn't gone after all.

As goose hunters know, the bird made a remarkable comeback. So much so that we shifted from preservation mode, to management mode, to concern over their growing status as a pest species. Abundant local birds were rounded up and shipped to municipalities in more southerly states, such as Nebraska, who wanted some for their areas. That stopped when they became pests there as well, soiling beaches and overpopulating golf courses. Incidentally, geese have to eat large quantities of vegetation because their digestive systems are relatively inefficient, which is why they defecate about every seven minutes!

We have explored several ways of controlling the burgeoning goose population. Geese harvested in Minnesota are given to local food banks. We have found nests and either addled the eggs (by shaking to disrupt the egg membranes) or oiled the eggs (halting the developing embryos). Most enjoyable to me is the metropolitan goose season with its liberal limits. Those first few days of the early season, with a bunch of naive geese flying around, make for some fun hunts!

A recent series of events from New York and Canada parallels our local experiences, with some East Coast twists. First, the goose population is around twenty-five thousand in the New York City area, whereas the target number is more like four thousand. The concerns about too many geese involve not tarnished beaches or golf course hazards but airline safety. We all remember the US Airways flight that ditched into New York's Hudson River because it collided with geese, and the captain's heroic efforts to save his crew and passengers. In fact no-goose zones with a seven-mile radius have been established around John F. Kennedy and LaGuardia airports in an attempt to keep planes safe during takeoffs and landings. More recently, in Ontario, a Porter Airlines flight struck three geese on its landing approach, damaging a propeller. But the geese have not read the rules about no-fly zones, and their large populations constitute a serious threat to aviation safety, pitting human against goose.

Goose-control personnel from the New York City Department of Environmental Protection rounded up excess geese, gassed them, and put them in a landfill. This did not set well with people from many walks of life. Hunters were dismayed that they were not allowed to cull the population, although I'm not sure a spread of decoys in Central Park would go over well. Actually, the outcry was more about using the meat and not wasting it.

This year officials from New York City arranged for the meat to be donated to food banks in Pennsylvania. Pennsylvania, but not New York, has a protocol for testing and processing goose meat. Testing for what, I'm not sure.

A hunter who raised concern, Jackson Landers, held a goose-cooking event at a Virginia winery, where the geese had been eating the grapes. Now that has to be pretty cool, given the adage that "you are what you eat." He helped stop the dumping of culled geese in landfills and made a strong case that the geese are good eating.

Canada, along with New York City, has not yet decided to allow Canada Geese to be offered at food shelves, as can be seen from the quote above by Gail Nyberg, although some cities and agencies are considering it. Nyberg also claimed that "serving geese at Toronto's Daily Bread food bank would be a 'nightmare' partly because the birds need to be plucked before cooking." Apparently, she is unaware of the fact that with a little practice you can breast out a goose in less than five minutes. I never pluck them. And I only save the breasts and thighs, but that's my preference (up to now; see below). She is also unaware of the fact that well-prepared (i.e., not overcooked) Canada Goose is excellent table fare. One thing that helps is a mechanical tenderizer made by Jaccard—it turns the toughest gander into very tender meat. Landers made an excellent point when he said that "eating [Canada Goose] would give greater meaning to the cull."

Opposition to the culling of geese in New York City comes from another, expected source. Sinikka Crosland, member of the group The Responsible Animal Care Society (TRACS), claimed that New York's plan to ship goose meat to food banks in Pennsylvania was "just an excuse to justify something totally unethical." Addressing that issue here doesn't seem necessary, unless you and anyone you know and care about intends to fly in or out of the New York City area. Also, the distinction between "unethical" and "totally unethical" escapes me.

There are nonlethal alternatives for goose management.

Mowing less frequently helps, as the geese like the new growth. Sprayable goose repellents are available, but they require frequent application. I still prefer letting hunters try to do the job, and Landers provided me extra incentive, namely, his recipe for goose thighs and drumsticks, called "goose rillettes with dried fruit mostarda." It's given below, and I'm looking forward to keeping the drumsticks and trying it.

At this point in time, we should view the recovery of the giant Canada Geese as a conservation success story. How we manage them from now on will be a challenge here and elsewhere. Scientific information plays a crucial role in these efforts. For example, goose remains from the engines of the downed US Airways flight were subjected to scientific testing to determine whether the geese were local or migratory. The tests involved analysis of feather isotopes, which provide in essence a bar code of where the goose was when it grew its feathers—if you're a bird, "your feathers are what you eat." The isotope profiles matched areas in Canada, leading some to argue that the birds that brought down the US Airways flight were migratory, not residents of New York City. However, we know that failed nesters and first-year nonbreeders fly north in summer to molt. So the birds that collided with the plane could have been "resident" geese returning from their molt migration, supporting the efforts to cull the local birds.

GOOSE RILLETTES WITH DRIED FRUIT MOSTARDA
From Jackson Lander and chef Dan Schleifer

Goose rillettes

4 goose hindquarters (leg and thigh)
3 ounces kosher salt
8 cloves garlic, crushed
1 tablespoon juniper berries, crushed
3 tablespoons black pepper, coarsely ground
4 cups rendered goose and/or duck fat (or fresh lard or neutral oils, not olive oil)

Rub the goose quarters with a mixture of the salt, garlic, juniper, and pepper, and allow to sit in the fridge for twenty-four hours.

Rinse the goose under cold, running water to remove the seasonings. Place the goose quarters in a shallow pan, and cover with the rendered fat. Cover the goose in foil, and place it in a 200°F oven for eight hours, or until the meat is fork tender.

Allow the goose to cool, and remove from the fat. Separate all of the meat from the bones and skin, and set the meat aside. Separate the fat from the drippings using a gravy separator.

Place the goose meat in a standing mixer, and mix on low, slowly adding the reserved drippings until the meat takes on a moist, spreadable texture. Add ¼ cup of the reserved fat, and salt and pepper to taste.

Dried fruit mostarda

1 cup each (coarsely chopped): golden raisins, dried figs, dried apricots
1 tablespoon mustard seeds
¼ cup of sugar
½ cup of cider vinegar
½ cup of water
½ teaspoon salt

Mix all ingredients in a small sauce pan, and bring to a simmer over low heat. Simmer for ten to fifteen minutes, or until all liquid has been absorbed. Add additional sugar or vinegar to taste. Remove from heat, and allow to cool.

To serve

Spread a tablespoon of room-temperature rillette over a toasted slice of baguette, and garnish with ½ teaspoon of cool mostarda.

48

CATS OUTDOORS AND NATIVE BIRDS
AN UNNATURAL MIX

To say that letting cats roam outdoors is controversial may be the biggest understatement since Noah (reputedly) remarked, "It looks like rain." Many have written about the detrimental effects on native birds caused by house cats that are allowed outside. Pure and simple, letting your cat go outside is ecological pollution. Our birds did not evolve with a feline predator of that size and ability in the ecosystem. It is not a just fight.

Now to be fair, not all cats are alike. Some stay inside, some go outside and do nothing, some go outside and kill in their yard, some go outside and roam and kill, and some are truly feral, big-time killers. What, you say, all cats don't kill? Well, it is true.

Kittens that were taken from their mothers before they were weaned tend to be incompetent hunters—I had one. The few times she was outside, her "stalks" consisted of flattening herself on the ground and trying to advance across open grass, teeth chattering loudly, toward a robin who was watching her from the get-go. She was never successful.

Well, I should qualify that. When we lived in Baton Rouge, Louisiana, she did once return with a dead mockingbird, one of the common local birds. I immediately recognized it, however, as the flattened roadkill that had been just outside our driveway for the past week. I guess that was her answer to being ridiculed and called incompetent. But what about serious data?

For many years, we have had only the studies of Stanley Temple, retired from the University of Wisconsin-Madison, to provide actual data on the effects of cats on wild birds (and native rodents). They radio-collared cats and followed them, estimating that upwards of seven million birds were killed annually by feral cats in Wisconsin alone, and that's just rural cats.

Still proponents of letting cats outside debated the numbers and claimed that they are made up by researchers, whom they consider to be hardened cat haters. I remember one case where a "study," done in Britain, I think, found that cats let outside did not kill wild birds—at least none were brought home. This finding was taken as counterevidence to the scientific evidence that cats kill a lot of native birds. But the real explanation was that cats had already killed most of the birds in the area, so of course none were left for cats to catch and eat!

Still, it was fair to say that although anyone who knows cats knows that many of them kill native birds and rodents, the scientific evidence was slim. That is changing.

A study published by Anne Balogh and colleagues in the *Journal of Ornithology* reveals in detail the devastating effects of cats on native birds. These authors studied one of our familiar native birds, the Gray Catbird, at three localities in the Washington, D.C., area. The study areas were residential, containing significant areas of vegetation, mostly landscaping. They located sixty-eight catbird nests and followed the fates of the eggs and young.

The researchers fitted sixty-nine nestlings and fledglings with radio transmitters. In one of the three localities, the birds actually had decent nest success, especially for an urban area. The area also had fewer cats. The other two sites were found to be population "sinks," meaning that without input from neighboring areas, the local population would go extinct, or figuratively, down the drain.

The local populations became sinks because of cats. Nearly 50 percent of predation events were a result of cats eating catbirds (irony noted). I found this passage in their article to be particularly revealing: "As such, domestic cats are likely able to intensively monitor, locate, and hunt inexperienced juvenile birds." The authors called roaming cats "subsidized predators," which brings to light the fact that these cats are not subject to starvation, disease, or competition from other cats. Now, perhaps, "ecological pollution" will mean more to those who let their cats outside. It isn't a fair fight, indeed.

Many cat lovers believe it is the "right" of cats to roam outside and kill our native wildlife. I like cats (indoors) but do not understand this lapse in logic. To let your cat outdoors, assuming it is a killer of wildlife (not all are good birders), means that you don't like native birds.

The University of Nebraska-Lincoln extension service recently issued a report on the problem of feral cats. The report noted that "a pair of breeding cats and their offspring can produce four hundred thousand cats in seven years under ideal conditions, assuming none die." Kind of staggering numbers, but they came from the Humane Society! This report also points out that although proponents of cats outdoors believe that well-fed cats do not eat birds, this is utterly wrong—they continue killing even if not hungry.

Many people also have a perception (unwarranted hope) that cats kill animals like domestic mice, rats, and pigeons, which would be doing us a service. This is countered by the fact that in a California study, 67 percent of rodents, 95 percent of birds, and 100 percent of lizards brought home by cats were native. In fact, house mice were more common in areas with cats than without them.

Feral cats kill not only nongame but game birds as well, like young pheasants and ducks. These cats have a high frequency of disease and usually have fleas and lice. They have shorter life spans than indoor cats, by at least a factor of two. Given all this knowledge, one wonders what it will take for people to keep their cats indoors.

It is not fair to our native birds to let them be eaten or injured by a house cat. I am especially disturbed when a stray cat catches a mother bird in my yard, leaving the babies to starve. Also, once the parents are eliminated, the young become so hungry that they beg incessantly, well more than usual, which allows predators, like squirrels and raccoons, that otherwise might not have found the nest to find it and dispatch the helpless young. Come on, cat lov-ers, don't you get it?

The Balogh study is welcome news in this debate because it provides solid data on cat depredation on wild birds. It shows how an entire local population can be changed from a self-sustaining population to a "sink" that cannot sustain itself via natural repro-duction. And it's the cats, pure and simple. We should be better stewards of our environment. Just like you don't dump your trash out your back door, neither should you open it for your cat.

49

FIVE MILLION U.S. RESIDENTS DON'T SEE THE PROBLEM WITH THEIR CAT KILLING JUST ONE BIRD A DAY

I admit to adapting this title from an article in a well-known satiri-cal newspaper named after a vegetable. But I had to get your atten-tion, because hasn't the cat-wildlife issue been visited enough times? Many have pointed out the problems caused by irrespon-sible cat owners who let their cats outdoors to kill native birds, rodents, and other creatures. It is pretty common knowledge that cats kill around a million birds a day in the United States. I liken letting cats outdoors to ecological pollution. Mostly, we need to bolster our rejection of Trap-Neuter-Return (TNR) programs, where cats are sterilized and deliberately released in outside enclo-sures or natural areas.

In a 2010 article in the prestigious journal *Conservation Biol-ogy*, Christopher Lepczyk and colleagues addressed the issue of cat depredation on wildlife, with some new or at least underappre-ciated twists. What impressed me most was that instead of just reviewing the problem, they made some concrete suggestions for action. I summarize them below and add some comments.

Conservation biologists, wildlife ecologists, and the like should have open dialogues with the animal welfare, sheltering, veterinary, and public-health communities.

Why? People who advocate letting cats outdoors either don't get it or are in denial. Or like the title of this essay says, their cat "only kills a bird or two a day." Ninety million cats in the United States and many let outside to roam—do the math.

Cats are the "leading vector of rabies among domestic animals." Also, cats that are trapped, neutered, and released have "shorter lives, higher levels of parasites and disease, and generally poorer health," all while they are out in the environment killing native animals. So, when TNR is presented as the "humane" alternative to reducing cat numbers, don't be fooled.

The wildlife and conservation communities should advocate for enforcement of requiring licenses for cats, decreasing unwanted breeding of pet cats through mandatory or subsidized spaying and neutering, and requiring cats to be kept under their owners' control at all times when outdoors. In many cases, local ordinances requiring owners to keep their domestic animals under control are enforced for dogs, but not for cats.

OK, this is a tough one but in my opinion the most important. People let their cats outside because they suffer no consequences for doing so. The problem is that most communities do not want to take on the issue of enforcing a leash law for cats. The argument is that a cat ordinance is "unenforceable." After all, if your three-year-old kid opens the door and the cat runs out, should you be cited? What about farmers who have cats in their barns for rodent control? They're likely to scoff at the idea that they should reduce or eliminate their barn cats because they are killing native birds. And the alternative, to poison rodents in barns, isn't necessarily better.

We can't win all the battles. But we need to change the culture of letting cats outside. We should make it clear that if a neighbor's cat visits our property, we will call animal control and have it removed, or at least file a complaint. Some will say that enforcing a cat ordinance will cost too much. The proper response is, why should the "blame" be put on those who make the call? Why not focus the blame for the increased cost of enforcement on those who let their cats out? After a while cat owners will realize that people will call enforcement if their cats are out and that they will be fined. If a road-killed cat has a collar with identification, the owner could be fined for letting it out.

> *Releasing cats into the wild and supporting feral cat colonies could be considered as a violation of the Migratory Bird Treaty Act and the Endangered Species Act.*

Great point. It is against state and federal law to cause the death of any bird in the United States except for introduced species (like house sparrows, starlings, pigeons) and legally taken game species. If your cat is out killing birds, you're responsible and are in violation of the laws that protect birds (not just "migratory ones," as the name of the act unfortunately suggests). The authors state that "it may become incumbent upon us to take legal action against colonies and colony managers, particularly in areas that provide habitat for migratory birds or endangered species." This would be fair turnabout.

Some readers will remember the recent case in Texas where a bird lover shot a cat that was hunting the threatened Piping Plover. He was charged and went to court, although the jury failed to reach a decision, and the judge declared a mistrial. I think justice was done.

> *It should be illegal to maintain cat colonies on public lands.*

Well, that's a no-brainer. All local communities should have an ordinance to this effect. Any misuse of public lands for this purpose should be promptly exposed and eliminated. I wonder what Floridians would think if there were areas where people could release their no-longer-wanted pythons?

We need to move away from the prevailing view that depredation of individual wildlife species does not matter as long as their populations are intact. If we are to consider cats from an individualistic viewpoint, then the same argument must be made for wild animals.

This is an excellent point. People, especially cat owners, are deeply concerned about their individual cat but not an individual chickadee or warbler. Heck, I had two cats for eighteen years, even though my wife, whom I married when the cats were five years old, was allergic to them. My cats were family pets, and we were as "attached" to them as pet owners get, so we worried about their health and dealt with all the vet bills necessary to keep them healthy. But they didn't go outside. The authors are advocating here a shift in thinking to the individual birds that cats kill. Cat lovers seem a bit hypocritical. Why should we not call attention to the nesting mother Song Sparrow who is snatched alive by a cat, killed after a period of "play," and eaten (or often not) while she was en route to her nest to feed her young (that now likely will starve)? Why is its life any different than that of a cat?

In conclusion, biologists do not often venture into the realm of making real-life recommendations, and I commend the authors for doing so. Whether it precipitates any action remains to be seen. I think we should take a more proactive stance toward cats that are allowed outside unsupervised. However, we need to be compassionate, because people's emotions do run deep when it comes to their family pets. Just consider the controversy a few years ago caused when a Wisconsin hunter suggested we shoot trespassing cats (again, this is legal in some states if the cat is feral or wild). Did it make many cat owners want to change their behavior and keep their cats indoors? We don't need hype. Instead, we have all of the biology on our side. Cats outdoors and unsupervised are ecological pollutants, which is easy to understand, and I hope someday relatively easy to fix.

50

CATS ON BIRDS
A MORE INSIDIOUS SIDE

One might think that the authors who titled their scientific paper "Urban Bird Declines and the Fear of Cats," published in *Animal Conservation* in late 2007, intended the title to be somewhat tongue-in-cheek. Indeed, the title does mislead in the sense that birds, to our knowledge, do not "fear" predators in the same way that we might feel fear if being stalked by a grizzly. But the authors were serious and bring up an important issue regarding the cat-bird debate that I think has been overlooked.

First, let's be clear about the natural history aspects of this issue. Yes, cats kill birds and rodents, and that is what they did before they were domesticated. So, it seems "natural" to a lot of people. However, be equally clear about this: our birds did not evolve in an ecosystem with a huge population of feline predators of that size and ability. Neither did they naturally encounter pesticides or pollution. So, cats are nothing short of "ecological pollution" when they are allowed to run outdoors and kill native wildlife.

House cats kill a million birds a day on average in the United States. This number seems high until you consider that there are close to ninety million cats in the United States. Many of these are allowed to roam, and especially in the southern parts of the United States they can be feral for most of the year and do tremendous damage. There are even misguided programs that promote releasing spayed and neutered cats into nature. Work by Stanley Temple's lab in Wisconsin estimated that at least seven million birds were killed by domestic cats each year in Wisconsin alone (and the upper-end estimate was over two hundred million). But there are naysayers. Some websites challenge the data from Temple's lab and say that the estimates are too high (the fact that there are zero scientific data to the contrary is apparently irrelevant) and cats don't do as much damage as commonly stated. But could there be more "damage" than direct killing of birds?

Enter the "fear of cats" from the title of the paper. The three authors of the study focused on England, where the number of cats has increased from four million to eight million in the past thirty years. The authors noted that in some places in England the documented effects of cat depredation on birds are relatively low. So, that would reinforce the notion that cats don't do as much damage as some people think. But there's a twist or two.

First, scientists have found that predators (e.g., cats) have an indirect, sublethal effect (fear) on prey species (e.g., native birds). In areas with high predator density, which one could argue is everywhere because cats add to the predator community, aspects like reproductive success declines and higher levels of stress lead to early death. For example, species that normally nest relatively near the ground may choose higher sites to avoid cats, and these higher nesting sites may be either less safe or less available. So, even if a cat does not directly kill a bird, it can have serious negative consequences on bird populations owing to lower numbers of young produced and higher death rates. That is pretty insidious, and considering that at least in England the ratio of cats to birds ranges from 35:1 to a low of 1.5:1, depending on the species of bird, birds have a lot of cats to be afraid of!

Although the authors of the study have no direct information on the "fear" factor, they used models to predict its effect. They concluded that given current cat densities in England, one could see a 95 percent decline in bird populations owing simply to the fear factor. The results make sense.

In some areas of England, cats kill relatively few birds, which leads some to believe the "cat menace" is overstated. But if you find an area with lots of cats, chances are you'll find many times fewer birds than you'd find in areas without them. In such an area, if you go out and track cats, you'll see that they catch relatively few birds. However, there are relatively few birds to catch in the first place, because they were mostly already caught and destroyed! As the authors note, "low-predation rates simply reflect low-prey numbers." So the data that show cats catching only a few birds are potentially very misleading—they can't catch what ain't there!

In summary, the authors concluded that cats can affect our native birds both directly by killing or severely wounding them and also indirectly by lowering reproductive rates through a fear factor. This puts a new twist on the suggestion by some that it is OK to let cats outside if they wear a bell or a plastic "apron." The new point is that simply by being where they shouldn't, they can have a negative impact on our native birds.

51

SOME WE LOVE, OTHERS NOT SO MUCH

Invasive species often threaten the ecosystems on which our native species depend. We mount vigorous campaigns against the spread of carp (of various kinds), purple loosestrife, spiny water-flea, Eurasian watermilfoil, emerald ash borer, buckthorn, and zebra mussels, to name but a few. These highly successful non-native species do indeed threaten a variety of native plants and animals, and we are working hard to control or eliminate them. We wonder whether the decline in the Mille Lacs walleye fishery is not at least in part due to zebra mussels, introduced and spread by the very people who enjoy the fishery the most.

And what more disgusting introduced creature can you imagine than the cockroach? Although there is a native roach in Minnesota (ironically, the Pennsylvania wood roach), the roaches most of us know, like the American and German, are introduced. Now, of course, roaches are fascinating biologically, but even I find stepping on them disgusting (and, admittedly, pleasing at the same time). I wonder if PETA supporters like roaches or they step on them like everyone else? Dislike of roaches may be some of the only common ground between hunters and nonhunters.

But our opinions on exotics are fickle, to say the least. Some exotics we absolutely love. Take honeybees for example. They are not a native species, but I enjoy honey immensely. They've always been in my environment, and honey has been a staple in my diet. I've conveniently forgiven their existence here in exchange for their delightful secretions. That would, however, change if our honey bees behaved like the "Africanized" versions.

The Minnesota Department of Natural Resources stocks rainbow trout in some Minnesota lakes. I fish for them and consider them outstanding eating. And they are incredibly fun to catch. They are not native to our lakes, which probably explains why my favorite bait is an entire (nonnative) night crawler with three pieces of (nonnative) corn. Do these trout damage the native fishery? Apparently not, as trout are apparently not evicting native fish species (except perhaps tullibee) filling the deepwater niche. In fact, native northern pike like stocked trout as much as I do.

What fascinates me about the biology of invasive species is why they are sometimes so successful. One might think that our environment is filled to capacity, ecologically. But the fact that some introduced species do so well either means that they do so at the expense of native species, or they are occupying "vacant niches." I can think of at least two types of vacant niches, natural ones and ones created by us. I think we have created some new niches that are filled more readily by exotics with preexisting adaptations than through the much slower evolutionary process of natural selection and speciation.

One of our best-known and loved exotics is the introduced Ring-necked Pheasant. Pheasant lore has become so integrated into our culture that we forget the bird is not native to North America. Annual pheasant hunts rank up there with deer camps in popularity. The pheasants came on the scene with a bang (pardon the pun). According to the Minnesota Department of Natural Resources, by 1931 Minnesotans were harvesting one million of the estimated four million pheasants in the state, less than fifteen years after introductions of a few thousand birds (in nearly all counties). Clearly the bird found a climate and food and nesting sources to its liking in its new home. And the feeling was mutual.

My demure, quiet, and reserved grandmother used to shoot

pheasants out the second-story window of her house in the western suburbs of the Twin Cities with a .22 single-shot rifle during hard times. My grandfather on my other side sold insurance in western Minnesota and used to have a 12 gauge in the backseat of his car when traveling from farm to farm meeting clients. During the Depression he would shoot a bird along the road if he saw one (he also told my mother that she was eating "swamp rabbit" when in fact it was muskrat).

The success of the pheasant raises some interesting biological questions. Did pheasants fill a naturally vacant niche, one that we created, or are they infringing and pushing native species out of their homes?

A study in Illinois found that about 5 percent of Greater Prairie-Chicken nests were parasitized by hen pheasants. In addition, roosters are known to physically dominate Greater Prairie-Chickens, including chasing dominant males off leks. Others have reported Ring-necked Pheasants parasitizing Lesser Prairie-Chicken nests in southwestern Kansas, although only three of seventy-five prairie-chicken nests were parasitized, and the few pheasant chicks that hatched apparently didn't survive.

Thus, Ring-necked Pheasants have seemingly had limited impact on native species. I suspect that the rise of pheasants and the demise of prairie grouse were a function of how we treated the landscape, making a niche for pheasants and removing one for grouse. Still, our attempts to recover Lesser and Greater Prairie-Chickens may be hampered by nest parasitism by pheasants. Of course, pheasants have become so engrained in our society that I cannot imagine a movement to eradicate pheasants from the landscape. But they are exotics.

Lastly, I cannot resist taking a shot at a long-standing pet peeve of mine in the category of exotic species. Lawns. I detest them. I detest the chemicals used to support lawns. I detest the pride and attention some give to lawns as status symbols and chuckle at the disgust some show at my "lawn." Why is it that we decry the loss of native species from our environment, rail against zebra mussels, and then return home early from a fishing trip to fertilize and mow the lawn? There ought to be a law against non-native lawns!

We moved into our house about twenty years ago. Its property has a lot of large oak trees, and the lawn would have rated a C minus at the time. In the ensuing twenty years, I have mowed it, but done nothing else, and it has slipped to a D. It is now a delightful mix of dandelions, creeping Charlie, mosses, lichens, and who knows what else. I refuse to encourage my lawn by using fertilizer, herbicides, or pesticides. Amazingly, some kind of grass still persists. My neighbor has an excellently manicured traditional lawn and, to his credit, minimal grass and a lot of shrubs. I try to mow my dandelions before they go to seed and blow across the road.

I look at the birds on lawns. I often have all five species of migrant thrushes foraging in my yard on their way north in the spring. Lawns treated with herbicides or pesticides usually have none. I also look at some local lakes choked with weeds (often introduced) and am amused that local residents wonder how this occurs. I can at least say that no fertilizers from my "lawn" were in the runoff that led to these aquatic blooms.

We are all hypocrites about exotic species to one degree or another. I find it interesting to ponder how some exotics rate our unbridled enthusiasm and others, outright disgust. I guess depending on the exotic in question, some people see the doughnut, and others see the hole. And yes, I realize that you're glad you don't live across from me.

52

RICO, THE CIRCUS, AND CONFLICTS BETWEEN HUNTERS AND NONHUNTERS

Oddly, people who strive to see animals killed ethically and humanely, namely, hunters, aren't always friends with the Humane

Society of the United States (HSUS), who, it would seem from its name, has a similar goal. Many people think that this organization runs animal shelters (a good thing), but we are also keenly aware that they disapprove of hunting. So, the seemingly common bond of humane treatment of animals is actually a fracture point.

Facts are not always front and center when pursuing an agenda. For example, the Humane Society disapproves of dove hunting, and one can read on their web page: "Mourning doves are the traditional bird of peace and a beloved backyard songbird. . . . Hunters kill more doves each year—more than twenty million— than any other animal in the country." Doves are not songbirds, which include species like robins and sparrows. It's like saying that a Pileated Woodpecker is a kind of sparrow. Their second statement is correct; doves are the most commonly killed game animal in the United States. But they go on to use scare tactics like, "Many hunters don't bother to retrieve the dead or wounded birds." Is that so? Where are the statistics on that? The statement's veracity apparently doesn't matter to the HSUS, because they think enough people will assume that if it's written, it's true. The HSUS also suggests that "American kestrels, sharp-shinned hawks, and other federally protected birds look like doves and can be shot by mistake." I would wager that more are killed by collisions with vehicles than accidental shootings, so their concern is not based on data.

The HSUS and other animal rights groups play on the emotions of people lacking sufficient background to know when information is not factual. Mourning doves are not endangered, although their population has decreased slightly over the past forty years. I'm sure some people are horrified to learn that the doves that nest in their yard are shot by hunters, but I also wager that quite a few would change their mind if they ever ate one. Given the proper negative emotional prodding, people lacking direct culinary experience with doves will send money to fund campaigns to ban dove hunting. Some of these same people, how- ever, think nothing of killing mice, running over snakes, squashing a roach, or flushing a spider. We clearly have a price structure on the value of animal's lives, and the HSUS preys on this fact. As you'll learn below, it's a profitable strategy.

I have over the years thought that the HSUS played a role in rescuing abandoned animals. I know a lot of people who have gotten fine pets from shelters. But this is not what the HSUS does. A watchdog group (Humanewatch.org) recently reviewed HSUS tax returns, which showed that they give less than 1 percent of their budget to pet shelters, put more money into lobbying than pet-shelter grants, and contribute more to their pension plan than to shelters. The HSUS spent almost ninety times more money on fund-raising than it spent on pet-shelter grants.

What, then, does HSUS do? In 2010 they spent $3,600,000 on lobbying and $47,000,000 on fund-raising–related costs. They had 636 employees, including 29 who earned more than $100,000. They lead large-scale legal assaults on not only hunters but farmers and ranchers, and even the circus industry. Yes, the circus industry. In fact, this is the main focus of my article (and be patient, I'll eventually get to what RICO stands for!).

The Friends of Animals (later merged into the HSUS), the American Society for the Prevention of Cruelty to Animals (ASPCA), and the Animal Welfare Institute (AWI) sued Ringling Bros. and their parent company, Feld Entertainment, in 2000 alleging mistreatment of elephants in circuses. You might be surprised to learn that the basis of the suit was that Asian elephants, those used in the circuses, are actually protected under the U.S. Endangered Species Act (ESA)! The United States accepts the worldwide Convention on the International Trade in Endangered Species (CITES) and extends protection in the United States to species endangered elsewhere. If a poacher is successful at exporting an illegally taken endangered species, it cannot be imported into the United States. That's good.

The ESA prohibits "take" of endangered species, and the suit by the animal rights groups contended that Ringling Bros. harmed elephants by using "a bullhook" (three-foot-long rod made of wood or fiberglass with a metal hook and a metal point on its end) and chaining them on hard surfaces (and trains). The main witness, Tom Rider, was a former employee of Ringling Bros., whose job was to take care of the elephants. He said that he formed an emotional bond with the elephants and claimed to have suffered emotional distress over their treatment, leading him to quit his job. In legal terms, he reputedly sustained "aesthetic and emotional injuries."

After nine years in court, Judge Emmet G. Sullivan dismissed the case, saying that Rider's testimony lacked credibility. Rider contradicted himself so many times, the judge said, that his testimony was "pulverized" by cross-examination. The judge noted that after Rider quit cleaning up after elephants, for the next nine years his only source of income was derived from animal advocacy organizations or their sponsors, who paid him to be a witness. I could go on for pages about this, but if you're curious about how the legal system is forced to waste time and resources, read the final opinion at https://ecf.dcd.uscourts.gov/cgi-bin/show_public_doc?2003cv2006-559.

The lawyers for Feld Entertainment learned a lot about animal rights groups, including the HSUS, during this lengthy and costly trial. They filed a countersuit against the HSUS and two of their corporate attorneys, three other animal rights groups, and a Washington, D.C., law firm. The Feld suit claims that these groups engaged in bribery, fraud, malicious prosecution, obstruction of justice, and money laundering.

This brings me to RICO (finally, thanks for waiting). What, exactly, does it stand for? I suspect that most readers are as unaware as I was that it stands for the Racketeer Influenced and Corrupt Organizations Act. The Feld suit contends that the HSUS and its followers are racketeers! The HSUS knowingly supported Rider by making direct payments to him to testify, which in effect is buying a witness's testimony. The Feld lawyers are also suing to recover the legal costs that defending their company accumulated over nine years (about $20 million).

According to Jim Matthews, writing for the Outdoor News Service, "animal rights groups have cost the farming and ranching industry jobs and raised the price of products we all buy every day. They are behind the efforts to ban sport hunting across the nation. They have forced state wildlife and fishery agencies to waste countless millions of dollars on lawsuits." Matthews goes on to say that these lawsuits could cripple animal rights groups, noting that Judge Sullivan's ruling "all but set the stage for a class-action RICO lawsuit against HSUS for misrepresenting itself in its fundraising campaigns across the nation. This future lawsuit could easily bankrupt HSUS and put it out of business—and send some of its top executives to prison."

This is a sad state of affairs for our country. We need to accept that different people have different standards for their personal behaviors. If you do not wish to hunt, don't overlook the fact that it's a time-honored, legal activity and is enjoyed by others. Hunting is in our heritage, and the argument that we don't need to hunt to eat has no value. The meat from a grocery or butcher that you eat was killed by someone for you. I choose to humanely kill game animals that my family enjoys eating. It's my right, and others should respect it. If it's not your cup of tea, I respect that as well, and won't demand that you become a hunter.

I object to the scare tactics used by some animal rights advocacy groups, and I would like to see these organizations better policed and punished for making false claims. I have no qualms with ensuring that animals are treated humanely, or as humanely as possible when used in tests that will help people. I am troubled, nonetheless, that even if these groups lose lawsuits and are forced to pay millions, those dollars will come from people who genuinely thought they were supporting worthwhile causes like local animal shelters.

ANIMAL INTELLIGENCE

53

A NEW RESPECT FOR PORCUPINE QUILLS

I own pointing dogs, which I let range far afield, and when we're hunting, several major concerns loom. One is that they'll get caught in a trap. Second, they'll find a skunk that they can't help investigate (been there, done that). Lastly, the dog will corner a porcupine and get a snoot full of quills (fortunately, not yet). We have all heard plenty of trip-ending stories about dogs and porcupine encounters, for good reason—the porcupine is a formidable opponent! Our local porcupine has about thirty thousand quills, which are actually modified hairs that are reinforced with keratin, a tough structural protein. Contrary to myth, the porcupine cannot throw quills at predators; they are only released upon direct contact.

My attention was drawn to a 2012 article titled "Microstructured Barbs on the North American Porcupine Quill Enable Easy Tissue Penetration and Difficult Removal," by Woo Cho and colleagues in the *Proceedings of the National Academy of Sciences*. My first reaction was, "Gee, haven't we known for a long time that the quills have barbs like a fish hook that hold the quills in, and it is these barbs that result in painful extraction even if you cut the ends and 'let the air out?'" But Cho went many steps beyond this and actually figured out how and why the quills are so effective. I read on.

Cho and his group examined porcupine quills under extremely high magnification. Quills have two distinct regions: a conical black tip that contains a layer of microscopic backward-facing barbs on its surface, and a cylindrical white base that contains smooth scalelike structures. The barbs overlap slightly, and there is a one- to five-micrometer space between the tip of each barb and the quill shaft. The size of the barbs becomes larger farther from the apex of the tip.

Most features of organisms, like porcupine quills, are the results of long periods of natural selection, making them better and better adapted to their particular functions. Of course, evolution via natural selection works within competing constraints, as we'll see. If a structure gets too good at doing one thing and it has multiple functions, then it may need to balance competing demands and not be as good as it could for a particular function. Compromise, then, is an evolutionary strategy.

Imagine that you wanted to make a porcupine quill from scratch. You would want it to have characteristics that would make it the optimal weapon. Your quill should provide excellent penetration, that is, maximal capacity to penetrate the skin and muscle of your adversary. Get it in deep is your number one priority. But then you would want it to stay there, and so you would want the quill to be difficult to remove. These two qualities, penetration and removal, might be oppositely opposed. For example, if the backward barbs were too big, they might impede penetration. If they were too small, the barb might be too easily removed. Somewhere in between must be just right.

Now for the fun point—who says research has to be boring? How would you "test" the value of quills for these two functions, given that using them on yourself or getting lab mates to volunteer seems rather unlikely? The researchers plunged quills into pig or chicken skin and muscle and measured the "penetration force" and the "pullout force." To have a basis for comparison, they carefully sanded off the barbs on some quills, making them smooth and needlelike. From others, they removed barbs from different sections, to see where the maximum effects were localized.

I had guessed that a quill shaped like a needle the nurse uses to give vaccines would be better at penetrating than one with barbs like a porcupine quill, and that the opposing needs of penetration and removal would sort of hamstring the quill's penetrating ability. But I was wrong! Cho found, using the same force, that barbed quills penetrate deeper and better than smooth ones. Further experimentation showed that the barbs within four millimeters of the tip help most with insertion. The reason is that the tissue is stretched and deformed by "high stress concentrations" near the barbs, and these "reduce the need to deform the entire circumference of tissue surrounding the quill, consequently reducing the penetration force." OK, that's a bit academic for me.

The authors pointed out that the concept of stress concentration has been used to design knife blades. Compared with straight-edged blades, serrated blades cut tissue more efficiently by localizing strain in the meat's surface at points on the tips of serrations. The strain concentration causes the tissue to fail with a lower input force. Consequently, serrated blades provide cleaner cuts with minimal deformation of the tissue. So, the barbs surrounding the quill act like serrations on a knife blade and make it easier to insert.

After learning that barbs helped, and not hindered, quill penetration, I figured that they wouldn't help hold the quill in. But at least that part of our intuition was correct. It takes much more force to remove a barbed quill than a smooth one. They found that tissue fibers interlock under the barbs, especially near the tip (maximum point of insertion). When the quill is pulled out, the barbs also bend, making extraction harder. They said that "barbs within a 4-mm barbed region at the apex of the quill work independently to minimize penetration force and cooperatively to maximize pullout from tissue."

At this point, one might wonder if studying porcupine quills is kind of frivolous. I mean, getting paid to figure this out seems like good work if you can get it. But Cho and his group had some loftier goals. They not only compared quills of differing structure but compared them to hypodermic needles used by physicians and explored the "development of a medical needle to achieve reduced penetration force" by fabricating a "prototypic hypodermic needle with microscopic barbs." Their prototype behaved the same as barbed porcupine quills, and presumably their keeping the barbs microscopic minimized the pullout force. This is a good example of having to solve competing problems: you want your needle to penetrate more cleanly and with less force but not be as hard to remove as a porcupine quill. Their work could actually lead to inoculations that would be less painful.

Many of our most convenient products were inspired from study of evolution's success stories. For example, study of cockleburs led to the development of Velcro. By studying the results of millennia of natural selection on porcupine quills, these folks learned a lot about what we only knew superficially (if not painfully). How natural selection has solved the opposing actions of penetration and pullout is fascinating. But rather than acquiring this knowledge for purely academic reasons of limited value to humans, the investigators concluded that "mimicking the porcupine quill should be useful for biomedical applications including local anesthesia, abscess drainage, vascular tunneling, and trocar placement in addition to the development of mechanically interlocking tissue adhesives." Maybe we should have a little more respect for porcupines after all!

OUTFOXED AGAIN
FOXES USE BUILT-IN RANGE FINDERS!

We tend to view ourselves as the ultimate observers. We can experiment, watch, record, and interpret what animals do in the real world, and we're pretty sure they don't have the same capabilities. Our confidence in our abilities was shaken, however, when we figured out that birds see in the ultraviolet spectrum. We now know that wild turkeys, for example, are communicating with each other via a channel we don't subscribe to, via UV-reflecting patches on their feathers. Actually, if you washed your old hunting clothes for years with the wrong detergent, it actually brightened their UV reflectance, making you stand out like a beacon. Oops.

Another surprise, albeit fairly old now, was that many animals make use of the earth's magnetic field. Birds, for example, use the magnetic field as a compass during migration. Cows and deer orient magnetic north when they are not stressed or being chased, and winter beds of deer are also oriented north–south. That means they can sense the earth's magnetic field and consistently orient north–south. Scientists aren't sure why they do this, but a possible reason is that when these social animals are dispersed by a predator, they can regroup easier if they all head in the same direction—magnetic north.

These results make one wonder if many animals use magnetic directional information in other ways we are unaware of. Hynek Burda and colleagues from the Czech Republic have now reported that the red fox senses and uses the earth's magnetic field in an unexpected way.

I have seen hunting foxes a few times, and they exhibit a behavior called "mousing." The fox detects a prey species, such as a mouse, approaches, sets itself, and then jumps high in the air and arcs downward on the mouse from above. Before the jump, foxes perk up their ears and tilt their heads back and forth, apparently using their sense of hearing to get an exact fix on the mouse. What's magnetic about this?

You would expect that if you watched one hundred attacks, they would be spread out over 360 degrees with the direction depending on circumstances like wind, vegetation, where the fox was when it detected a mouse, and so on. However, researchers had noticed before that fox attacks were not directionally random. That's suspicious. So the Czech scientists had twenty-three wildlife observers record fox attacks and the direction they oriented when they executed an attack. The observers kept track of whether the jumps were in low or high grass and whether they were successful.

In low cover there was no deviation from random. However, 75 percent of all successful attacks in high cover were centered about 20 degrees clockwise of magnetic north. That's clearly not random.

One could think of many factors that might explain these results other than magnetic orientation. However, the authors pointed out that the observations were recorded at all hours of the day, different times during the year, under cloudy and sunny conditions, with different wind directions, and by different observers. That erased many of the factors I thought might "explain away" their observation.

The authors instead concluded that "directional heading has a profound effect on hunting success under conditions in which visual information is not available to augment auditory cues." That is, when they have detected something in tall grass and they can't see it, they need an attack guidance system.

That still leaves me scratching my head, wondering what the heck it would matter to the fox whether he "moused" from the north or the east. How could this information possibly be useful? The authors suggested that "mousing red foxes may use the magnetic field as a 'range finder' or targeting system to measure distance to its prey and thus increase the accuracy of predatory attacks."

When foxes tilt their heads sideways, their ears are at different heights above the ground and therefore different distances from the noise the mouse is making. This allows them to home in more precisely on the mouse because the noises reach the ears at slightly different times (owls use the same mechanism). But that's apparently not enough.

Next the authors suggest that "a fox that approaches an unseen prey along a northward compass bearing could estimate the distance of its prey by moving forward until the sound source is in a fixed relationship to the magnetic field, e.g. it coincides with the inclination of the magnetic field. This would consistently place the fox at a fixed distance from its prey, allowing it to attack using a highly stereotyped leap."

Daniel Cressey blogged about it this way. "Think of a laser pointer attached to you that always points slightly downwards in the same direction. Now think of some object on the ground. If you walk toward the object until the laser spot is on top of it you know that object is a set distance away." That is, foxes have an onboard "targeting system" to measure distance to prey and thus increase the accuracy of their mousing attempts.

I'm still thrilled by the new things we learn about nature that have been in front of us for centuries. Heck, I even checked to see if I took a northward route to the fridge. However, being directionally challenged, I gave up, being glad I don't have to sense magnetic north to find the leftovers.

55

HOW DO GROUND-NESTING GROUSE EVER BREED SUCCESSFULLY?
AN OILY SUBJECT

Spring and summer are not the favorite times of year for my English setter and my drahthaars. Every day they check me out in the morning when I come down the stairs to see if, by any chance, I'm wearing hunting clothes. After this wardrobe check, when they realize that we're not going bird hunting, they begin their ritualistic dances that precede their a.m. treat. Gotta love their ability to switch to Plan B and still be your best friend.

But I got to thinking about what they'd do if they were afield in the spring or early summer (which is illegal). I'm pretty confident that they don't miss many birds when they're hunting. A few molecules of scent from a bird that passes their noses gets their attention, and anything more than a few gets a solid point. So, I'd bet that they could find just about any hen sitting on a nest, be it a Ruffed Grouse, a Sharp-tailed Grouse, a Hungarian Partridge, a woodcock, or a Ring-necked Pheasant. After all, the hen sits for considerable periods, and this has to lead to a good downwind scent cone that surely my dogs would catch. They might miss a rapidly moving bird (like a running pheasant) but surely not a sitting hen.

That in fact is why you are not supposed to be out running your dogs during the bird nesting season. Even if they don't catch a bird, they can disturb it. When undisturbed, most incubating hens leave their nests quietly, skulking away to avoid detection. But a sudden flush could well attract the notice of a predator, who later finds the nest, or the hen wouldn't have the opportunity to cover the eggs (to conceal and keep them warm).

Or am I overestimating my dogs' ability? It is folklore in some places that hunting dogs cannot find a quail or woodcock hen sitting on a nest. Maybe they just weren't well trained. But what about coyotes, foxes, and wolves, who depend on their noses for food? Surely they have even better scenting ability than a hunting dog? If so, how could a hen on a nest ever be successful, as during the three-week incubation period, wouldn't you expect a coyote, raccoon, or fox to at least once cross a scent cone coming from the sitting hen?

A bit about the scent cone, which we should know from our dogs. A predator is more likely to encounter a scent cone when it is long and linear. This occurs when airflow is smooth and not turbulent. If air is turbulent and of high velocity, the scent cone is hard to locate, as the scent molecules are dispersed and don't form a direct line back to the nest. Also a scent cone can be lifted over a predator's nose by updrafts. Now, if these theoretical ideas are correct, birds can "hide" from olfactory predators by putting their nests where updrafts, turbulence, and faster winds predominate.

Another key to hiding from olfactory predators comes from wax or oil. All birds have a gland (one of the few skin glands birds have) located on top of their tail at its base. The gland goes by various names, including preen gland or uropygial gland, or the Pope's nose. It produces an oil that birds get onto their bills by squeezing the gland with their beaks; they then preen their feathers, transferring the oil to the feathers. To get it on their heads, they use their feet.

For a long time we have known that these oily or waxy secretions are how birds waterproof their feathers. These oils allow ducks to float. They allow a bird to fly through a rainstorm. Without the oils the feathers are not waterproof, and the birds become waterlogged. A duck would drown, and a robin would likely perish flying through the rain. The reason is that without waterproofing the bird becomes thoroughly wet, and if the temperature is cold, the bird dies. If the temperature is not cold, the bird won't be able to fly, making it vulnerable until its feathers dry out. Keeping feathers waterproofed is a life-and-death deal for a bird.

But waterproofing has a cost, namely, that the oils can be smelled by predators. The chemical composition of these oils has been studied. Two basic types are known, for simplicity, monoesters and diesters, which differ in their characteristics. Remember, what you smell are particles in the air. Diesters are heavier and therefore less volatile than monoesters; so diesters make a scent cone harder to detect because the scent cone "sinks." Why two types of these preen oils? We know that some birds switch from monoesters to diesters during the breeding season. Could this have a nest-concealing function?

Researchers in the Netherlands did some experiments, using a trained six-year-old German shepherd. They trained the dog to find tubes that contained monoesters or diesters (versus blanks), and in hundreds of trials, the dog did not make a single mistake. They then tested monoesters versus diesters, and the dog had much more trouble finding diesters, especially at lower concentrations. So, for a nesting hen, the chemical switch to diesters also functions as an olfactory camouflage.

The results from the Dutch researchers suggest that a switch to a diester preen oil may explain why hunting dogs struggle to find woodcock or quail hens on their nests. Of course, then you would ask, why don't they use diester oils year-round? They may be more costly to produce or be less efficient as waterproofing. They may affect the way the feathers reflect light and hence change the visual appearance of the bird to other birds. And when a bird doesn't have a nest to protect, it can simply fly away if it is detected by a scent-sniffing predator. So the bird wants the best preen oil possible in the nonnesting season even if it is easier to smell.

There have been some other hints that birds behave in ways to mediate the scents they produce. For example, a study of Sharp-tailed Grouse showed that when they are loafing, they stand in places with greater updrafts, wind velocities, and atmospheric turbulence, making them harder to detect by scent-seeking canids. In a study of Greater Sage-Grouse, researchers determined that nests were located in a way that inhibited both visual and olfactory predators. Raptors and ravens will eat eggs and young, and concealing the nest from them is also important. Many sage-grouse nesting areas have both visual and olfactory predators, so the birds balance the threats by putting nests in places that are equally difficult for the two types of predators, thereby splitting the difference. The camouflage plumage of nesting hens also helps avoid predators.

The preen gland secretions are fascinating. Evolved as a way to waterproof plumage, these secretions cost the birds by giving away their locations to olfactory predators. To mitigate this cost, natural selection favored a switch to a less detectable but still functional oil (diester) in the nesting season, to provide a scent elimination system (and here we thought that we thought of this first). This then seems to be a primary way that ground-nesting grouse avoid losing their nests to predators like coyotes, foxes, and wolves. And other mammalian nest predators, like skunks, raccoons, weasels, have good senses of smell, and even an elk—yes, an elk—was recorded on video eating sage-grouse eggs. And remember the deer that smelled you last year, in spite of your scent-eliminating shower and your scent-free clothing, hat, boots, and gum? Even whitetails will eat eggs and young if they find a nest. All these predators make the probability of a successful nest rather low, and thus, there is intense natural selection on nesting hens to avoid detection.

Of course, nature is never static. If you're a predator thwarted by birds' chemical switch to diesters during the nesting season, the onus is on you to be better at detecting them. The more we learn, the more we realize that nature is an ever-evolving arms race between predator and prey. Darwin would have loved it.

56

OUR CHICKADEES ARE SMARTER THAN THEIRS

The Black-capped Chickadee is one of eastern North America's most familiar birds. We see them in many places all year long, from deer stands to duck blinds to bird feeders, from the forests in the far north to woodlots in the prairie. It is one of the first birds to announce the upcoming spring, when in January the males begin giving their two-toned whistles (sounding like "phee-beee"), which are used to set up and defend territories.

One of the most striking and impressive things about these little birds, which weigh four-tenths of an ounce, is their winter tolerance. While we huddle around a hot cup of coffee first thing on a subzero morning, the chickadees are at the feeder—active, interacting, seemingly oblivious to the cold that could kill you or me in short order without proper attire.

So, recalling that they were here long before we started putting out feeders, we ask, how do they make it through the winter? Well, they can spend the night in a cavity, out of the wind, maybe huddling with some close friends, as this will help conserve energy during the night. But during the day, it's eat eat eat. They have to eat constantly to survive, and very long without food, maybe only a day or two, will likely spell doom. And studies have shown that carrying around a lot of fat is not a good idea, as it reduces aerodynamic efficiency and makes the bird easier for a predator to catch.

There aren't many bugs flying around my place in the winter. None, actually. But chickadees hunt for small insect larvae overwintering in the bark of trees, eat seeds still left on weeds, and maybe visit a feeder or two. But these sources probably don't provide enough food to keep the birds from freezing to death.

A trick of chickadees is food caching. When times are good, chickadees store food in tree bark, which they can retrieve during winter when food is harder to find. The problem is finding it again when you're hungry. You have to remember where the heck you put that seed, was it here or over there, or maybe halfway up the first branch on the north side of that big oak in the backyard with that English setter? Birds remember?

Yes, chickadees have memory. They either remember specific storage places or just have an idea of where they would have stored a seed last fall, like having a search image. I must admit that as I am losing more and more memory in my late fifties, I try to put things in logical places so that I can find them later after only two or three tries (after first blaming my wife, incorrectly, for moving some item). Chickadees, by finding and eating food that they stored in the fall, supplement what they find, and make it through the winter. But they need a good memory, which I guess makes it fortunate that chickadees aren't big drinkers.

A 2009 scientific paper by Tim Roth and his colleagues described their studies of chickadee brains, and I have to warn you that the methods are a bit dramatic. Under permit from state and federal agencies, they went to several places around the country, captured chickadees, and then humanely killed them. Next, they did some special tricks and sectioned their brains so they could measure the areas associated with memory (specifically the hippocampus, a part of the brain involved in spatial memory). They were testing whether chickadees that live in harsher environments (more cold and snow and shorter days) had bigger areas of the brain associated with memory than did chickadees that live in milder places, because to survive the winter, they had to remember more food-caching spots.

They collected birds at several places, including Maine, Washington, Iowa, and my backyard in Grant, Washington County, Minnesota. Each of the locations is at about the same latitude, so they "controlled" for day length, otherwise birds in areas with longer days might have smaller brains because they had more time during the longer days to find food. That's good science.

I am happy to report that Minnesota chickadees have the biggest brain areas associated with memory! Well, they at least had the highest relative number of hippocampal neurons, and that's good enough for me to declare this proof that our chickadees are the smartest (the Maine birds were pretty smart too). They have to be to survive our crazy winters.

So, like our strong women, good-looking men, and our smart kids, Minnesota chickadees are above average, too!

57

NECK-DEEP IN GUANO
A RECENT HISTORY OF CHIMNEY SWIFTS

Everything we do, did, and will do is framed in the context of time. But given the universality of time, we have a surprisingly poor grasp of it. I know, usually, what I'm doing at the moment, but I don't have to go far back in time, say, yesterday, for what I did to become blurred (if not forgotten). I do occasionally remember that such and such a place has changed dramatically over the years. I am sure that winters have been less severe, but apart from that, I wouldn't be much help in reconstructing even the recent history of our environment. A recent scientific study by Joseph Nocera and colleagues titled "Historical Pesticide Applications Coincided with an Altered Diet of Aerially Foraging Insectivorous Chimney Swifts" discovered a piece of environmental history that has a lot of interest to all of us.

Chimney Swifts are often described as cigars with wings. Indeed they have a cigar-shaped body and long narrow wings. They fly rapidly and erratically while they catch insects on the wing. They often nest in chimneys, hence their common name, and in fact the first one for my "yard list" came after we moved into our home in early March and I found the mummified remains of one while cleaning out the fireplace chimney! (I suppose purists would argue that this doesn't count, but my list is rather loosely argued.)

Chimney Swifts often nest in colonies in larger chimneys, such as those associated with schools or factories. During the nesting season they go in and out all day, feeding babies in nests glued to the sides of the chimney. They often gather above the chimneys toward dusk, all swirling around, making their entrance to their nests or roosts to spend the night. While they are inside the chimneys, their droppings and bits of regurgitated (indigestible) insect remains fall to the bottom, where they accumulate over time. It turns out that there's more to guano than meets the eye (or nose). The minute bits of undigested insects can be identified under a microscope, and the guano itself can be surveyed for various chemicals, such as pesticides. Given that the top of the guano stack is recent and the bottom is old, one can develop a chronological record of what they have eaten over time and develop a history of pesticides in the environment, which are taken up by the insects they ate. So, swift guano allows one to follow the health of the ecosystem over time.

Some insect-eating birds that feed on the wing (e.g., swallows, flycatchers) have experienced serious population declines in the past few decades, and the Chimney Swift is an example. One can see a major decline at the national level and a comparable one in Minnesota. An obvious question is, what is behind these declines? Is it a natural phenomenon unrelated to human activity, or has our alteration of the environment played a role? Candidate factors include deterioration of conditions on the wintering grounds, reduction of preferred prey types on breeding grounds, loss of nest sites, pesticides, and climate change, to name a few.

Nocera and colleagues discovered a two-meter-deep guano stack (which they referred to as a "geochronologically dated deposit of guano and egested insect remains"—don't you love how scientists describe feces) in a chimney in Kingston, Ontario. The chimney was capped in 1992 but contained a fifty-year record of Chimney Swifts, during which time their population dropped by over 90 percent. The chimney was home to thousands of swifts. The authors of the study analyzed the guano and insect remains for changes in diet and the link between shifts in prey abundance and historical pesticide deposition. They took slices of the guano stack from top to bottom and used radioisotopes to date the slices. The authors plotted the relative abundance of beetles, true bugs (things like cicadas, aphids, plant hoppers, leafhoppers), and DDT. Pretty sophisticated given the subject "matter"!

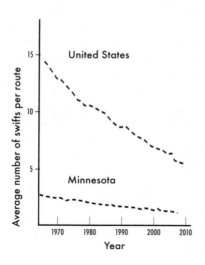

Examination of the insect remains showed a major shift in the swift's diets between the 1940s and early 1950s. During this time, the representation of "true bugs" increased, whereas the percentage of beetle remains decreased. Basically, if there are fewer

Plot of the average number of Chimney Swifts detected per survey from 1966 to 2010 as part of the U.S. Geological Survey's Breeding Bird Survey.

flying beetles, swifts switch to true bugs, which are not as palatable (they give off chemical defense odors—think of stink bugs) or nutritious. (Try saying "swifts switch" really fast one hundred times.) This trend reversed from the late 1950s through the 1970s, when swifts consumed relatively more beetles.

The authors noted a pretty striking relationship between DDT in the sample and numbers of beetles. In general, beetles are hit pretty hard by DDT, and their populations show a clear negative response. The true bugs are good at rapidly evolving adaptations to pesticides like DDT, and they do not seem to be affected nearly as much as beetles. I'm not sure why this is, but given their ability to create noxious substances (again, think stink bugs), perhaps they have a physiology suited to "defusing" pesticides.

Although DDT use in the United States stopped by the 1970s, the record of swifts shows that it persists in the environment for a long time afterwards. In fact, there was a slight upturn in the amount of DDT and a corresponding decline in beetles consumed. Thus, swifts were experiencing "residual exposure" to DDT, and it continued to affect their prey. The relationship between DDT concentrations and swifts' use of beetles and true bugs is pretty hard to dismiss as a random correlation. Those who dispute the role of DDT as an environmental contaminant have some tough explaining to do here.

Still assigning a particular cause to the decline in swift populations is dif-

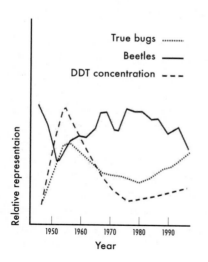

Temporal pattern of occurrence of true bugs, beetles, and DDT in samples of Chimney Swift guano from Ontario. From Nocera et al.

ficult. The authors point out that swifts might have encountered DDT on their wintering grounds, where it was used much longer than in the United States, and not obtained it from local insects. Habitat loss may also have contributed to the decline of these birds. Alternatively, their numbers may have been depressed so far that it will take a longer time for them to recover. However, this assumes that beetle populations will reach sufficient densities to support populations of swifts as large as they once were.

We have no reason to expect that effects of DDT were limited to birds that forage "on the wing." The study of swifts teaches us a poignant lesson about the long-term effects of chemicals that we use to control our environment. Nocera and colleagues provide a good reminder to all of us who care about the environment that being good stewards of the land and water requires vigilance, care, and continual scientific input. Even I am amused, however, by what scientists can do when given enough guano.

SHAKE, RATTLE, AND SPRAY, DOGGIE STYLE

My English setter is great at finding birds and holding a point. I once watched a bird walk under him at a game farm, and he didn't break point. He once pointed a wounded, but live, bird under four inches of snow with no visible signs that anything was there. When I hand signal to him to search for birds in places he doesn't naturally head, he looks at me as if to say, "If you know where the birds are, why do you need a dog?"

There is one thing my setter doesn't like, and that's water. He's afraid of thunder (even from a stereo) and rain, and he will cross a stream only under the most dire threats. I've never bothered to send him on a water retrieve. One exception to his dislike of water is the kiddie pool at my local hunt club, in which on a warm day he loves to plop down to cool off. But then his expertise with water shows up again when he stands up and shakes, still in the pool, and then lies down again. And, of course, we all have noticed that despite a nearly infinite number of places dogs can go to politely shake themselves dry, they prefer to shake themselves off as close to us as possible. I'm not sure why they feel compelled to share this moment. Maybe they're wondering why you didn't get in the pool too, and just want to share.

I'd never really thought about the process of dogs shaking water out of their fur. It seems instinctive and just one of the things that dogs do. But after a moment, you realize that it's a universal instinctive behavior, one they obviously inherited from their wild ancestors. Probably any behavior we see in an animal at least originally had a basis in natural selection that resulted in improved survival of those that do it, relative to those that don't.

Basics. Why, in the first place, is a dog wet? OK, seems a bit obvious, but the reason is surface tension that keeps the water attached to the dog's hair. Why might being wet pose a problem to a dog? Is it more than just an inconvenience? A wet dog is a lot heavier than a dry one. So, one advantage in rapidly shedding water would be to get back to escape, pursuit, or cruising speed ASAP. Water in the fur also reduces its insulating capacity, and when it freezes, it has even worse. So having a rapid drying mechanism can be seen as a basic adaptation that evolved to compensate for having hair as insulation, hair that can get wet and retain water. But is there anything special about this process of shedding water? I recently saw a scientific article that actually studied how mammals like dogs shed water from their fur via that familiar shaking we've all seen a hundred times before.

Researchers at Georgia Tech led by Andrew Dickerson decided to study this process of shaking water from fur. They began with a mathematical model and tested it by filming shaking animals with a slow-motion video camera. The shake starts with a twist of the head, followed by several oscillations of the body as the energy wave travels toward the tail end. A dog can shed 50 percent of the water in its fur in less than a second (and, I'm probably wrong, but my perception is that my setter is able to direct all of it at me).

The metric for assessing how fast an animal oscillates during drying is cycles per second, or hertz. The researchers found that larger animals shook at around four hertz, but a mouse, closer to 30 hertz. Incidentally, our bodies cannot accomplish this feat, but if you want, get down on all fours after a shower and try shaking like your dog—you'll still need your towel and possibly a chiropractor.

There's some state-fair logic to the results. For those of you longing for your bygone days in physics class, the point is to generate maximal centripetal force on the water in a dog's fur. Centripetal force varies with the distance from the center of the body, explaining why large animals can shake more slowly and still generate enough force to break the surface tension. You can think of it this way: if you sit in the center of a merry-go-round, you experience less force on your body than you do at the edge. So if you were a drop of water, you'd be spun off way faster at the edge of the merry-go-round than at the center. Bigger merry-go-rounds can spin more slowly.

But as in many scientific studies, the researchers found a glitch: considering the animal's radius alone didn't fully explain the cycling frequencies. In particular, the animals seemed to shake at a slower rate than predicted. The answer lies in their skin. Many mammals, like dogs, have quite loose skin, and this means that the skin can move farther during each shake cycle than the underlying body, generating even more force to dislodge the water than the body alone could generate.

OK, you might be raising an eyebrow and wondering why someone would research this. Dickerson said, "our lab is basically interested in bio-inspired design and understanding mechanisms that happen in nature. And the original thought was that we could use what we find in this research to design a more efficient washing machine or design something that we haven't even thought about yet." Hey, even I know that we are interested in how fast our washing machines cycle. I love the notion of "bio-inspired design."

Even though I will still be annoyed when my setter decides to spray me, at least now I have a better understanding of the origins and potential functions of this behavior—and, maybe, even a slightly greater degree of appreciation. Apparently he's using his quick-spin cycle so he can get back to hunting and keep up his torrid pace, as his ancestors did, not to please a human but to survive and be the fittest. I wonder why we don't often stop to think more deeply about things we observe every day. I'm glad Dickerson did.

59

DRAHTHAAR FOLLIES

Every dog owner has some stories about events in his dog's life that at first greatly angered him and then in retrospect were just plain humorous. Here's one of mine.

On a recent trip to Texas we were bow hunting hogs at a ranch, and the guide killed a rattlesnake. Although I don't condone the random killing of snakes, my son had just discovered a rattler in the stick blind he was hunting from, and the guide would prefer that his clients not get bitten on his hunts! The snake's skin was really bright, and it had probably just shed. If you've never studied a rattler skin close up, you might not appreciate how incredibly beautiful they are.

We decided to skin it out and brought the skin home frozen. A lot of places in Texas have mounted rattler skins over doorways, and I decided that such a mount would be a nice addition to our family room. I bought a tanning kit online, cleaned the skin (several hours of fleshing), tanned it, pinned it out, and let it dry. I kept it in the garage for a month or so—as the tanning solution is defined as "toxic"—and it looked great. My son and I visited a local fabric store, where I thought we should get an award for being the first males ever to enter. We bought a mounting board and some felt fabric.

I laid it all out on the dining room table (my wife is very accommodating, and we eat in the kitchen a lot), put the dried rattler skin on top, and after a couple of weeks decided to rise from a Sunday afternoon nap and mount it permanently. So I sauntered into the dining room, rubbed my eyes, and stared in disbelief. It was gone! I looked under the table, thinking it had slipped off. Or maybe I put it somewhere else and just forgot? No, I was sure the last time I looked at the table, it was there. A call to my wife, who was out shopping, got me a snippy reply: "No, I haven't touched the snake skin and have no idea where it is."

My dogs. I figured that maybe my drahthaar or English setter had snatched it from the table, so we searched the house and then went outside to see if they had dragged it out, but no luck. I was then royally steamed and thought, well, if the stupid dog ate it and dies, too bad. After ten minutes or so, I thought about the fact that both are highly trained hunters and family pets, so I relented and took them both to an emergency vet clinic, which of course on a Sunday has triple normal fees.

Each dog was induced to vomit, and nothing came up. Now I really hoped I hadn't put it back in the garage or somewhere else. Home we went, me inside, the dogs into the yard. I watched the drahthaar do his "business," went out, and—sure enough—saw it was full of snake scales. The dog had eaten the five-foot tanned skin, rattles and all, like it was some sort of doggie treat. Fortunately the tanning solution had long since dried, and he showed no ill effects. But I was out $350 for the vet, many hours of prep work, and a really cool mount.

The dog checked out the dining room for a couple more weeks, just in case some new novel treat appeared (and a search for old food morsels is never beneath a dog). Apparently a five-foot, tanned rattler skin is a lot like the rawhide chews you can buy for your dog's enjoyment. I later wondered what he did when he got to the rattles but could only guess that the crunching sound was an added, if not interesting, bonus. Now, I guess, it would be a bad idea to hunt him in rattlesnake country! Once you've had tanned rattlesnake, there's no going back.

60

"TRASH BIRDS," THE LAW, AND AMAZING BIOLOGY

The Migratory Bird Treaty Act of 1918 made it illegal for any citizen to possess (or sell) any part of a bird, including a feather, egg (even an egg shell fragment), or nest, from any species, excluding the House (or English) Sparrow, European Starling, and Rock Dove (Pigeon). So if you are out on a frigid Minnesota winter afternoon and find a nest of a native, long since migrated bird and think it would make a nice wall decoration, stop. It is illegal for you to possess it. If you have a shed feather from a Red-tailed Hawk in your cap, you are breaking state and federal law.

This might sound a bit outrageous, almost legal frivolity. However, from an enforcement perspective it makes sense. How could a conservation officer determine whether you shot the bird and took a feather, or took the nest when the bird was still using it? The consequences are serious. If found guilty of possessing a protected bird, the maximum penalty is a fine of not more than $15,000, imprisonment for not more than six months, or both.

There are exceptions, however. A person associated with a museum or academic institution can apply for a scientific collecting permit. You must be working on a valid scientific study that requires collection of bird specimens, and you must justify the number you ask for. The specimens must be deposited in a scientific collection, such as the one I curate at the Bell Museum of Natural History at the University of Minnesota. If you work for a nature center or a school, you can apply for a salvage permit, allowing you to possess protected birds for display and educational purposes, the stipulation being that you cannot have caused the bird's death. If you find a dead bird and you can bring it to an accredited place, just make sure you do so fairly promptly. Permits go to public institutions; you cannot keep specimens in your home. At the Bell Museum, we get many donated birds each year, and most are valuable scientific additions to our research collection.

Other exceptions include game birds that you harvested during an open season with a valid license. Thus, you can mount your ducks or upland birds, but I would retain some written documentation. If you own a game farm and hawks are killing your birds, or a fish farm that attracts egrets, you might get a permit to control them, but you would likely be asked first to explore other options.

The last set of exceptions includes those species mentioned above, the sparrow, pigeon, and starling. These are introduced species, or exotics, for which there is no protection. They are generally considered pests, and they have caused quite a few documented ill effects. Ironically, other introduced species such as the Ring-necked Pheasant are now highly valued and protected.

The House Sparrow is, frankly, a nuisance. Aggressive and feisty, they often chase other birds from feeders. They often displace native birds from their nesting cavities. They have a high reproductive potential and can pull off several clutches a year. But they are pretty amazing creatures.

Released along the East Coast in 1850–52 as part of the we-miss-the-home-country attitude of new immigrants, they quickly spread throughout North America. By 1910 nearly all of the United States had been colonized. By 1951 they had reached Tierra del Fuego at the southern tip of South America. Because they were already accustomed to living in cities and just about anywhere else with food, they were great colonizers. They are great examples of at least two biological phenomena.

First, although they only colonized the continent recently, they already show variation in body size that conforms to Bergmann's rule. Bergmann's rule states that in warm-blooded vertebrates, individuals in northern populations tend to be bigger than those in southern, warmer climates. The reason has to do with the relationship between body surface area and volume. As you get bigger, your volume increases as a cube, but your surface area increases only as a square. So the bigger you get, the relatively less surface area per volume you have, and you lose less heat to the outside environment. House Sparrow body sizes have become larger in cold areas in less than one hundred generations. This is a really fast response. Imagine what could be accomplished over millions of years.

A second interesting thing we learn from House Sparrows concerns animal behavior. Most animal societies have a well-structured set of dominance relationships. Among male House Sparrows, the amount of black in the "bib," or throat patch, varies considerably and is a good predictor of who will beat up on whom. If an older, black-bibbed alpha male meets a young small-bibbed male, there is no question in either of their minds ("minds" probably gives sparrows too much credit) who's superior; you have to know your place in society! If you're a girl House Sparrow (you would not be sporting a black bib), you would preferentially mate with a male with the best possible genes, and the one with the biggest bib has proven by living a long time that his genes are likely superior.

The other two introduced species can boast a similar set of abilities to colonize new areas. Pigeons, originally from Europe, are now established worldwide. They have few natural enemies, although squirrels likely prey on the young (squab), and introduced Peregrine Falcons like to knock them out of the sky, catching them before they hit the ground and eating them. Obviously, we need more peregrines. And pigeons have a place in American music—who can forget the song by Tom Lehrer "Poisoning Pigeons in the Park."

I once thought that if I could only find a place where there were starlings and no one around, and if I had my twenty gauge, I could obtain some specimens for the museum. One day long ago in western Minnesota, I found an abandoned farm with some star-lings on the defunct power lines leading to the deserted house. I got out, took my shotgun out of the case, walked toward them, stopped, broke open the gun and loaded it, and then got back in the car. They left as soon as they realized I was heading toward them. (Yes, I unloaded and recased my gun.) Smart defines star-lings.

But one of the most fascinating things about starlings is their ability to fly in enormous, coordinated flocks in perfect synchrony. You can easily Google up some videos of big flocks of starlings, usually near dusk, where the entire group moves in unison, making high-speed intricate turns, and never, ever hitting each other. I cringe at an air show when there are three planes fly-ing together. But thousands of starlings flying wing tip to wing tip is indeed one of nature's wonders—perfectly choreographed, the entire flock moving in seemingly random directions but pur-posefully. This behavior has been observed for centuries, but no one was quite sure how they managed it. You never see two birds collide. Recently some physics-savvy scientists watched film of the flocks in slow motion and discovered how the starlings execute these amazing maneuvers.

Basically each starling follows a set of flight rules that involves the nearest six or seven other starlings, which coordi-nates their individual movements; that is, their flight radar keeps only the nearest six or seven birds on the screen, and they shift their own individual flight path accordingly. A starling near the edge of the group has a more extended set of birds on his flight radar, so all of the groups are interweaved. This way, the thou-sands of individuals in the entire flock behave like a coordinated wave.

Despite the amazing things that these introduced birds do, it would be better if they were doing them in their native ranges. Many of our native birds have suffered because of these introduced species. Once again, this brings home the extreme importance of stopping the spread of all exotics, from aquatic to terrestrial, even if some of the new immigrants do some cool things. But it doesn't hurt to learn a thing or two from these unwelcome newcomers.

61

THE DATING GAME, ANTELOPE STYLE

People have known for quite some time that children born from marriages between close relatives often have serious defects. We term this inbreeding. A prime example of the negative effects of inbreeding in people was revealed in a recent study of the Habsburgs, who ruled in Spain from 1516 to 1700. There were a series of Habsburgs kings, and to keep the position in the family, many marriages were arranged between uncles and nieces and between first cousins. For example, Charles's father, Philip IV, was the uncle of his mother, Mariana of Austria; his great-grandfather, Philip II, was also the uncle of his great-grandmother, Anna of Austria; and his grandmother, Maria Anna of Austria, was simultaneously his aunt. Family get-togethers must have been, um, interesting if not complicated.

One of the Habsburgs' hallmark physical abnormalities was a condition known as mandibular prognathism, or the Habsburg lip, resulting from the lower jaw growing faster than the upper jaw. Chewing was almost impossible. The Spanish dynasty ended with the death of King Charles II in 1700, who was physically and mentally disabled and childless despite two marriages.

The "extinction" of the Habsburgs was almost certainly a result of being highly inbred. We now understand that each of us carries some genetic variants that are bad (or "deleterious" in genetics terms). However, humans have two copies of every gene (except those on the Y chromosome), and for many bad genes to be visibly expressed, you have to have two copies. That is, if you have one bad gene, it can be masked (or it is recessive) by having a "good" copy on your other chromosome. You're then a "carrier." Close relatives, like siblings or first cousins, often share the same bad genes, and therefore marriage between them greatly increases the chances that a child will inherit two copies of the same bad gene. Hence, we have taboos against such marriages for very good reasons.

What about the rest of the animals? A scientific study by Stacey Dunn and colleagues from the University of Idaho published in the *Journal of Zoology* revealed that pronghorn antelope "know" about the negative consequences of inbreeding and "try" to avoid it. Their study was a cool blend of observation of pronghorns at the National Bison Range in Montana and genetics work in the lab. And it took a lot of time and effort to figure out!

First, some background. Once very abundant, pronghorns are now less common owing to habitat loss and fragmentation. So that in itself could be a potential problem. If a local population has few individuals and they don't recognize their relatives, an individual might mistakenly mate with another antelope that's too closely related for their offspring's good. In fact, in an earlier study following a population crash, inbreeding was observed in these pronghorns, and the effects included decreased fawn survival to weaning, low birth mass, short foot length, and poor body condition.

If you've hunted antelope, you know they are social and that there are "herd bucks" who have harems. However, what is probably not apparent to most of us is that females actively switch harems during the mating period, and so they can rate different males before choosing which one will sire their offspring. Yup, mate shopping and swapping. Females leave harems of males who lose fights with other males. Ouch. And as moose do, some females manage to get an external male to challenge the main guy to make sure he's still boss. Our pronghorn doe wants to make sure she's getting the best male, and she's not above calling the question! Actually, there's even more to it. Does almost always have twins, but about 10 percent of the time, each fawn has a different father. Apparently some females can't make a clear choice, and they split the difference.

Still, to my eye at least, it's not clear how a female would know whether she's related to another male. Maybe she recognizes siblings and uncles from personal experience or "smell." Given that herds of pronghorns are not like the last generations of the Habsburgs, they must somehow avoid inbreeding in general, if they have choices. How?

The researchers captured fawns and took a small tissue sample from an ear or collected fecal samples to provide material for genetic analysis (you can recover DNA from fecal material, although the fieldwork lacks glamour). Each fawn got a unique ear tag. They knew the genotypes of adult males and females in the population because they had captured them as fawns. This information allowed the researchers to ask two key questions: (1) Do does mate with males less related to themselves than the average male in the population? (2) How are does related to males in whose harems they "sample"?

The DNA technology is fairly straightforward (unless done by the LAPD). You genotype all the males, females, and young and determine how genetically similar they are to each other. The most genetically similar animals are likely closely related (i.e., fathers-sons, mothers-daughters, sisters, brothers), whereas at the other extreme, the least similar are likely distant cousins at best. By watching the females, the researchers could tell if does mated with males that were less genetically similar to themselves than a random male. This would mean that she was in effect avoiding inbreeding. If a doe mates with a male as genetically dissimilar to herself as possible, the chances of her young inheriting two copies of a bad gene are lowest.

This is exactly what was found. Female pronghorns mated with males that were less genetically similar to the female than females were to the average male in the population. Was this also true in the harems the females "sampled"? No. How genetically similar a female was to the herd buck had no relationship to whether she hung out in his harem. So she might in fact have been hanging out in a harem ruled by her uncle. But she did not ultimately mate with closely related males. Maybe she somehow figured out during her visit to a harem that the male was a close relative and hence didn't choose him as the sire of her kids.

Given their observations, the researchers concluded that does practiced inbreeding avoidance during the actual mating period, which for them is twenty-four to forty-eight hours, but not in the weeks leading up to the "event." Like any situation in nature, there were some mistakes. As I noted earlier, the same authors found in an earlier study that about 18 percent of the fawns were "moderately inbred" and showed the effects mentioned above. However, this might be a function of low population size and a shortage of available males. Thus, although inbreeding avoidance is a good thing, antelope have not perfected it. Apparently, they avoid it if possible. Too bad the Habsburgs didn't figure that out in time.

62

CAMOUFLAGE
ONE OF LIFE'S UNIVERSALS

Blending into your environment can take many forms. One dictionary defines camouflage as "the devices that animals use to blend into their environment in order to avoid being seen by predators or prey." Most hunters have a variety of camo clothing designed to blend into different habitats to conceal themselves from different game species. Camouflage is not always the squiggly patterns of grays and browns that we are used to. When hunting deer or turkeys from my ground blind, I wear a black shirt and mask to blend into the black background of the blind.

Examples of camouflage are widespread in nature. The coloration of hen pheasants and other ground-nesting birds allows them to blend into the nesting area. Many insects look like leaves or twigs. We could spend hours tallying up the ways that animals conceal themselves. But the point of this essay is to explore the question, when did animals start using camouflage? A 2012 article on the evolution of camouflage in insects by Ricardo Pérez de la Fuente and colleagues in the *Proceedings of the National Academy of Sciences* suggests it's been millions of years!

Insects and many other creatures are not only camouflaged by their coloration but by things they gather from the environment. Pérez de la Fuente and colleagues wrote about the green lacewing, one of about twelve hundred species of lacewings. Adults fly around, looking not unlike mayflies. But immature lacewings wander about on plants, eating other insects. To conceal themselves, larvae have special structures sticking out from their bodies that help them accumulate stuff, or trash, from the environment. The trash can be bits of plant material or shed exoskeletons from other insects. By adorning itself with debris, the lacewing blends in with the environment—it has a sort of mobile blind.

Pérez de la Fuente discovered an amazing immature lacewing in amber from Spain that lived in the Early Cretaceous, about 110 million years ago! On the top of the larva were special processes that formed a sort of basket, which they referred to as a "dense trash packet." They discovered that the basket contained parts of an ancient fern, known to be an early colonizer of fire-prone environments (the fires likely led to the amber resins that trapped our now-famous trash-hording immature lacewing). The authors speculated that the trash basket had at least three functions: camouflage, a defensive shield, and chemical defense. Certainly the trash provided mobile camouflage. Predatory insects, like "true bugs," have long mouthparts that they can stick into a lacewing's soft body; the authors suggested that the trash packet shielded the larva from such an attack. Lastly, a compound (phenols) in the ferns was speculated to act as a chemical deterrent—a sort of predator-away scent! So over 100 million years ago, this bug had camo down pat. What else do we know about camouflage?

The evolution of camouflage is not a once-and-done deal. Instead, much of nature is involved in a cyclical game of deception. Your species evolves a coloration that matches the background, and you avoid being eaten. However, predators, hungry and equally able to evolve, gain the ability to see your new look, finding you once again to be a tasty treat. Then you have to quickly evolve to another state that will fool the predator again, for at least the time being. We call these cycles "evolutionary arms races." They're everywhere you look. Think about why you don't get a flu shot just once instead of a different one every year—yes, humans and influenza viruses are in an ongoing arms race.

We can find examples of arms races at different stages. For example, it seems to me that a fawn would have zero chance of surviving, given that coyotes or wolves roam the area with their highly evolved senses of smell. There should be no deer left, but we know that's not true. Fawns produce almost no odor, which means that scent-tracking predators are "blind" to the fawn's location, and fawns blend in pretty well with their environment, so they are hidden from predators that rely only on vision. We can predict then that if wolves or coyotes are to become good fawn predators, they will need to evolve even keener senses of smell. This would then shift the pressure back to deer to evolve even less odor. And so on. I think it's fair to say that an arms race is on with deer currently in the lead, at least at the fawn stage.

Of course, nature is never simple, and not everything is designed to blend in to the background. In some cases, animals "want" to be seen. For example, many wasps have a bright and conspicuous black and yellow pattern that instead of concealing them sends a message to potential predators that they can deliver a nasty sting. In fact, many different wasp species have evolved to look similar to each other, a phenomenon known as Mullerian mimicry. The idea is that if you all look more or less alike, predators may avoid you because they've already learned the consequences of attacking a look-alike species. A strategy of mutual saturation of the prey landscape benefits all.

In addition, many harmless species have been adapted by natural selection to look like noxious species, getting protection under the coattails of the truly nasty species. This is called Batesian mimicry. My favorite examples are the flies that mimic bees. At many an outdoor sporting event, I've had a bee-mimicking fly land on my arm. I announce to bystanders that I'm going to catch it alive. I quickly cup my hand over it to the gasps of onlookers, who assume I'll be stung. But a quick study will teach you that the anatomy of flies and wasps is extremely different, so you look at head and body shape instead of the false-advertising coloration, and you avoid a nasty sting and potentially impress bystanders. The mimicry does apparently work in nature since so many flies are bee mimics, so not all predators must have figured it out, at least yet. But you can bet that the evolution of deception and countermeasures is ongoing.

Camouflage is widespread throughout the animal kingdom. Given that it is found in groups that have been distinct for millions of years, one should suspect that use of camouflage is an ancient invention by many different critters. And finding a 110-million-year-old fossilized larva caught in the act of using "trash" from its environment for concealment and protection puts a number on the minimum age at which camouflage evolved. I guess we shouldn't be too impressed with our own efforts at camouflage given that even insects invented it over one hundred million years ago!

ONE MORE CUP OF COFFEE

How many times have you been fishing or duck hunting and told yourself "one more cup of coffee and we'll call it quits"? We must have all done this once and then either hooked a fish or had ducks rain from the sky over our suddenly irresistible decoys. That must be what keeps the tradition going, but I have a hard time remembering when it last happened to me. Still, I do it, and I'm guessing a lot of others do too.

On a cold May Saturday morning my sixteen-year-old son, Chris, and I were in our ground blind waiting in ambush for a group of toms to come up a road, an event the landowner said was thirty to sixty minutes after sunrise each day, like clockwork. Yes, I know, it's called hunting and not shopping because you cannot predict, but we were still thinking this was a slam dunk about to happen. Zero dark thirty turned into 6:00, 6:30, 6:45, well past the guaranteed "appointed time," and we realized that not only had we not heard or seen a turkey, but it was a lot colder than we thought, and we were clock watching. We figured the birds had skirted us and were busily displaying in an adjacent sunny pasture. At about 7:15 I decided I'd get out of the blind and check the adjacent pastures to see if we had indeed been outsmarted. I didn't spot any turkeys, but as if to reinforce our doom that day, I did find the landowner's dog fifty yards from our blind, and he wanted to play. After ditching the dog, I returned to the blind and uttered, "Well, one more cup of coffee and we'll call it quits for today."

No sooner had I put down the thermos, then Chris whispered excitedly, "Here come the toms!" I peeked out of the blind, and sure enough three big toms were coming right toward us, not thirty yards away! I took a quick look at the long beards swaying as they walked and knew this was going to happen right then!

While parking my still-steaming coffee, I looked at Chris, and he seemed frozen, not by the cold but by the rapidly approaching birds. I had to spur him to action and hurry! The toms split up, one going on my side (my bow was not set up, as we had agreed that Chris got the first shot), the other two on Chris's side of the blind. The birds were walking fast, and the first one got by his shooting window. I realized I had to do something, so I picked up my slate call and gave a soft yelp when the second tom was broadside at twelve yards, stopping him in his tracks. His head stretched up, and I'm thinking, "SHOOT, SHOOT," but nothing from Chris. The reason I had such a good view of the tom was that Chris had to slide off his chair to his knees to get a shot off. As he did this, I couldn't see the tom anymore, but when Chris released the arrow, I knew he had connected.

From the same direction as the first three toms, several other toms headed our way, but they caught movement in the blind and hung up at fifty yards, even though they were gobbling in response to our calling. Checking in the direction of Chris's bird, I saw the first tom walking back toward us. Unbelievably, he had noticed our hen decoy and seemed interested!

Turkeys can be unpredictably dumb. The tom walked right toward our hen decoy, and at ten yards I got my shot. He ran off in the direction of Chris's bird, and that's where we found them both, five minutes later. What luck—two arrowed toms within a few minutes, after assuming the worst. Even more fortunate for us, as we got out of the blind and spotted our downed birds, we looked up the road in the direction the turkeys came from, and saw the landowner walking his dog home. Another couple of minutes and I'd have finished my coffee and we'd have gone home empty. You can be sure that the tradition of "one more cup and we'll leave" will continue to be a regular ritual!

POSTSCRIPT
CONFESSIONS OF A THREE-MINUTE OUTDOORSMAN

I THINK IT MUST BE TRUE that you cannot take the boy out of the man. Despite the heavy responsibilities that come with being adults and parents, we still find time to do the things we loved as children. My outdoors experience as a youth involved fishing. Although I spent nine years as a professor at Louisiana State University in Baton Rouge, where the state license plate read "Sportsman's paradise," I didn't wet a line or hunt any kind of game. When I returned to Minnesota, fishing was calling from my past. We had two young boys. I bought my first fishing boat. To help convince my wife, I announced that it was the "perfect boat for our family." It wasn't a year before this myth was exposed. With two young sons trying to cast hook-laden lures, places to duck and cover were too few and far between—we needed a bigger boat. "It *looked* like the perfect boat, only smaller," I had to confess, as we traded up. Fishing various local lakes and especially the St. Croix River became an addiction. We are now on boat number three.

I was also hooked on grouse hunting, and at the population peak in the mid-1990s, I could almost always bring home a limit. I hunted with others who had dogs, and before long we had the first of our highly trained and expensive hunting dogs. Of course hunting requires a small arsenal, and the common saying that "a man can't have too many guns" became a reality. Shotguns in different gauges, over-unders and semiautomatics, a large selection of shells of different shot sizes and pellet material (lead versus steel)—all became necessary tools. Teaching young boys to shoot requires youth-model guns. And a gun is not a gun is not a gun. A good grouse gun is almost automatically not the preferred gun for the duck blind. And you need to develop the knack of making these pronouncements with such confidence and conviction that your wife thinks you actually know what you're talking about.

A common pattern with newfound addictions is the need to keep them going year-round. A membership at a local hunting club provided an extended bird-hunting season. And with all those guns and an almost delirious joy in using them, what better than an addiction to sporting clays? When you hunt, shots are relatively few and far between. Shooting the sporting clays course guarantees fifty or more shots in a short time! You don't use just a couple of shells, but boxes of ammo. We formed our own sporting clays team at the club, which of course requires a lot of dedicated commitment and time. Involving the sons meant that getting a "house pass" was a given, as what mother could deny her sons a chance to go shooting with Dad? I tried not to let my wife catch me smirking too often as we saddled up and loaded the guns and ammo into the truck. But in reality, she didn't miss any of the agenda I thought was cleverly hidden.

Living with my mother in south Minneapolis, hunting of any kind, especially big game, was not part of my upbringing. Eating venison—something common to many Minnesotans—was also absent from our lives. A boyhood friend's dad who did hunt gave us some venison, which my mother dutifully took and lied about how happy she was to get it. She really had no intention of letting a piece of a deer pass her lips, but she knew if we didn't at least cook it, I'd tell my friend, who would tell his dad. She turned it into a petrified piece of dark matter that looked like a battered hockey puck, and it was no surprise that it had no taste and later required lots of dental floss. That was my view of deer hunting for nearly thirty years—it was a ruse to get your wives to let you hang out with the boys in the fall, drink beer, and shoot big guns.

Thirty years later, another acquaintance gave us another package of venison. Dutifully I accepted it with a response eerily similar to the one my mother had given my friend's dad. The difference was that my wife and I prefer our red meats to be very rare, and so what the heck, we tried it. Plus, a fine red wine, lacking from my mother's table, could be a worthy antidote. After a couple of bites, I sat stunned. It was one of the best things I had ever eaten. Venison even done to medium is rubbery and tastes terrible. Cooked rare, it is amazing. I had been misled, admittedly by myself, for three decades. I was hooked.

The next addiction was predictable. But I had to figure out how to bring venison to our table. Although for years I had collected hundreds of small birds for scientific collections and my research, I wasn't completely sure I could shoot a deer. Those big eyes, spotty fawns, maybe I'd hit a wall. If it was a wall, it was quickly scaled. A guy at the hunting club hunted deer with a bow, and he had one for sale. It suddenly dawned on me that a lot of deer were in our own yard. So I bought the bow. Not since being ten years old and shooting an arrow into the neighbor's birdhouse had I shot a bow. Time to learn a completely new skill.

Not sure how this would sit with the powers at home, I hatched a foolproof plan. I went to the local archery shop and bought a bow for our older son, who was ten. We had been pondering what to get him for Christmas, and I happily announced to my wife that I had solved the gift dilemma and had bought him a compound bow. "Well isn't that kind of dangerous?" was her response. "Oh, no," I said, "We'll join a local indoor archery league, and it will be another father-son bonding experience." She furled her brow and said, "Wait, but you don't have a bow." Time to fess up, and I had given this explanation some thought. I pointed out how busy we were with two young boys, rarely time to talk, almost like two taxis crossing in the night. I told her, "I just hadn't had time to tell you that I had bought this used bow and was intending to hunt deer with it." "Ah, I see," she said, "You wanted a bow, so you bought one for Chris for Christmas." Even I was a little shaken by how fast she figured that one out. However, involving a son is always an added benefit when asking forgiveness and not permission.

The addiction hit hard. Bow hunting season is three and a

half months in Minnesota. From mid-September to New Year's Eve, I spend nearly every day either on our property or a couple of other "spots"—which I learned you describe in general, not specific, terms, and no one should ask for more specific details. The unwritten rule is "find your own spots and don't ask me where mine are."

But three and a half months go by so, so quickly when you're doing something you love. Not necessarily to my wife. She does not see my being unavailable nearly every day at sundown or the dinner-fixing time as an unprecedented opportunity to try new recipes . . . I try to make up for it the rest of the year. If it weren't for the fact that our older son was born on January 1, I would think it's the worst day of the year, because there is no other day during the year after which there are more days before the next archery deer season opens. I actually look forward to January 2!

I had read about places in Texas where you can go and shoot arrows at hogs and exotic deer and sheep year-round. We don't have to put our bows away on January 1! We started a now-annual family trip to Texas over spring break to visit a ranch that stocks hogs and other critters. And not wanting to make the same mistake I made with venison, we bring back meat from hogs, sheep, and even goats. Smoked goat, boar, and ram are real treats.

I had it really bad. Invited by my wife's dad to Hawaii for the holidays, I was stuck. I like the Texas spring trip because I hate sitting on beaches, and bow hunting with my sons is a hundred times more enjoyable to me. What is there in Hawaii except lying on a beach and roasting like a beached whale? A quick Internet search found a bow-hunting place in Maui. Suddenly Hawaii seemed like the perfect place for the holidays. We brought home some axis deer and a ewe.

And then there was the banquet with the silent auction at the Minnesota Outdoor Heritage Alliance, where an African bow-hunting safari was being offered for the ridiculously low starting bid of $1,000. What an outrageous thought, and I couldn't possibly pull this one off. To get the bidding started, I offered $1,000.

I figured why not, I won't get it. I later noticed a very attractive woman talking excitedly on her cell phone and gesturing at the written description of the safari. Well, I figured, this is the beginning of my end in the bidding, so I'll up my bid to $1,500 to drive up the price for the good cause of supporting MOHA's goals. What could it hurt? I had mentioned to my wife that I was "bidding," and she mentioned something about how many glasses of wine I had drunk. Going to Africa was clearly over the top—there have to be some limits. I assured her there was no chance that I'd be committing those kinds of family resources to an African trip.

My older son and I returned from Africa with a bunch of nice trophies and memories that will certainly last my lifetime. He has since had trouble understanding that this "trip of a lifetime" was not to be an annual experience. And I now "owe" his younger brother a similar trip, which is currently under discussion. How awful it is that I have to be the common denominator in these once-in-a-lifetime trips. Incidentally, I "won" the African trip, as no one else made a bid, so the only two bids of the evening were mine, and yes, I outbid myself by $500. Apparently two addictions conspired together to eliminate common sense.

It doesn't end there. Several hot mid-August periods have seen a son and me sitting near a waterhole in South Dakota waiting for an antelope to come within bow range. We now have an annual upland bird and waterfowl hunting trip to central Manitoba. Even my wife likes this trip. For years, she declined our invitations, thinking that her role would be to cook and clean up, which even I couldn't bring myself to say was probably spot on. But she now loves the trip, which is out of cell phone contact, and she finds walking afield with the dogs and her "boys" a blast, even without a gun of her own. Then, for a few days of the spring, there's the turkey season. A whole drawer of stuff is set aside for this, as well as a bunch of turkey decoys stacked on top of a cabinet. The neighbor who introduced me to grouse hunting found a place we could hunt the "metro goose season," and I look forward to this as much as any hunt. One part of our garage is dedicated to storing goose decoys, which over the years seem to breed and multiply.

So the boy becomes a man, but the past doesn't die easy. For me, life is now much richer than it was when about all I did was work. I wish everyone at least some time where they can find happiness in things outdoors. Get addicted or readdicted—you won't regret it.

ROBERT M. ZINK is the Breckenridge Chair in Ornithology at the Bell Museum of Natural History and professor in the Department of Ecology, Evolution, and Behavior at the University of Minnesota. An avid hunter, fisherman, and bird-watcher, he has published more than 150 scientific articles and received the American Ornithologists' Union Brewster Medal in recognition of his research contributions.